Institut für Informatik
der Technischen Universität München

A Theory of Aspect Orientation

Jorge Alberto Fox Lozano

Vollständiger Abdruck der von der Fakultät für Informatik der Technischen Universität München zur Erlangung des akademischen Grades eines

Doktors der Naturwissenschaften (Dr. rer. nat.)

genehmigten Dissertation.

Vorsitzender: Univ.-Prof. Dr. Hans Michael Gerndt

Prüfer der Dissertation:

1. Univ.-Prof. Dr. Dr. h.c. Manfred Broy
2. Prof. Dr. ir. Mehmet Aksit
   Univ. of Twente / Niederlande

Die Dissertation wurde am 8.08.2007 bei der Technischen Universität München eingereicht und durch die Fakultät für Informatik am 11.12.2007 angenommen.

Gedruckt mit Unterstützung des Deutschen Akademischen Austauschdienstes

Bibliografische Information der Deutschen Nationalbibliothek

Die Deutsche Nationalbibliothek verzeichnet diese Publikation in der Deutschen Nationalbibliografie; detaillierte bibliografische Daten sind im Internet über http://dnb.d-nb.de abrufbar.

ISBN 978-3-8325-1856-1

Logos Verlag Berlin GmbH
Comeniushof, Gubener Str. 47,
10243 Berlin
Tel.: +49 030 42 85 10 90
Fax: +49 030 42 85 10 92
INTERNET: http://www.logos-verlag.de

Dedication

To my mother, who always stands for what she believes in. To my beloved sisters: Paty, Emma, Ana, and Mónica, whose light illuminates my path when needed. To my brother Fernando. This work is also dedicated to my father and to my brother Claudio, *in memoriam.*

In general, to all those resisting mainstream discretely or openly, for this is the only way to prevent social injustice.

Acknowledgements

Thanks foremost to Prof. Dr. Manfred Broy for the opportunity to investigate this subject. Prof. Broy was always ready to discuss the topic, explore it through examples, questions and provide valuable input. The intellectual freedom at his chair, his knowledge and experience enriched this work as well as the author's research and working capabilities.

Thanks to the German Academic Exchange Office (*Deutscher Akademischer Austauschdienst* -DAAD) for the generous financial and institutional support, especially to Ms. Hilde Mönch.

Prof. Mehmet Akşit exercised a guidance that can be compared to that of a lighthouse or navigational star; at the distance yet always prompt to share his expertise. Thanks.

A number of people shaped this dissertation in close collaboration or intensive discussions with the author. They provided an additional scientific guidance that is needed not only to achieve results, but also to recognize dead ends and avoid mistakes. We may not say this work is error free, but we have given it a thorough thought. Thanks a lot to: Dr. Jan Jürjens and Prof. Dominikus Herzberg.

In the same spirit, Dr. María Victoria Cengarle and Dr. Martin Wildmoser deserve special thanks as well informed "outsiders" ready to contribute their experience for a number of important related "aspects" of a dissertation.

Most colleagues at the chair contributed to this work by exchanging ideas at one stage or another. Two of them deserve a particular mention: Dr. Leonid Kof and Florian Deißenböck.

This work was also influenced through publication submissions to workshops and conferences. The author was sometimes successful, sometimes just ready to learn and try again. Thanks to the many blind reviewers. My appreciation to Awais Rashid for friendly comments on the contents of the fifth chapter.

Thanks to Marsha Chechik and Ed Brinksma for remarks on an early version of a "taxonomy of aspects". Chapter three progressed from that taxonomy.

Someone told me years ago that Sir Isaac Newton used to say "stand on the shoulders of giants." I just did.

Abstract

Aspect orientation is a software engineering technique that claims to provide an enhanced separation of concerns. The idea is that concerns that affect several modularization entities (i.e. crosscutting concerns) can be better managed by first identifying and then weaving them in their selected entities. This is supposed to improve modularization of software, facilitate reuse of code or software components and has even claimed to support software evolution. These claims appear to be well founded, though a critical evaluation is still missing. We do not aim at evaluating all these claims, instead of that we explore aspect orientation and try to establish what the *aspects* are, how these can be modeled and what use they may actually have.

Most work in aspect orientation provides enhanced modularization constructs at the programming and modeling levels. But in some cases the application of these techniques is independent of the problem itself. This means that the techniques for weaving at code or modeling level are in principle applicable to a number of problems without a clear criterion to answer questions such as, when should we actually use these techniques?

This leads to some of the questions we try to answer in this work: what do we consider an aspect, when do aspects arise, how do we deal with aspects, are aspects " crosscutting concerns", are aspects " views", are " services" aspects? Most of this are and remain open questions.

Our starting point is the following. The first notions of aspect orientation relate to crosscutting in code (the *bottom up* approach). We believe that aspect orientation can be better understood from an architectural perspective (the *top down* approach). We explored the above questions from a top down perspective, while keeping a clear reference to the programming level techniques that are representative of this approach.

Crosscutting is originated from the transformation of requirements into software design or specification. We have shown this at the hand of a case study with informal requirements in the form of use cases and its transformation into object oriented classes together with formal specifications in a lightweight formal language (Alloy). *Crosscutting* is defined as a requirement that is later expressed as some function or predicate (specification) affecting more than one modularization unit (e.g., class or component). This is in principle independent of the architectural style we choose, for instance: component oriented, object oriented, service oriented, etc. Though, it may be a consequence of it.

We argue that crosscutting results from the design decisions in the translation from requirements to design and later stages in software development. *Aspects* are defined as the representation of a crosscutting concern encapsulated in an additional modularization layer in a software model or code. Furthermore, based on the accepted classification of requirements, we identified two main kinds of aspect: *functional* and *non functional*. We focus on functional aspects and recall the notion of communication refinement to explain aspect weaving. Further on, we will demonstrate that weaving in functional aspect orientation can be explained by this notion. Indeed weaving in functional aspect orientation can be explained by this notion, granted that we consider the weaving function and the selection mehcanism independently. This means that communication refinement is not enough to explain the selection of the so called join points or the base elements into which the aspect will be weaved.

We will explain the use of our approach in the case of two aspects. These are two particular security aspects whose interaction we analyzed at the model level. We will introduce a composition model inspired on Composition Filters. Our model relates to the FOCUS formal theory. This theory allows for analyzing the (execution) history of communication channels in components. We will provide a formal model of composition filters and illustrate on a few well defined points of execution how this framework opens the door to analyze aspect composition and interaction issues.

In summary, we provide an explanation for crosscutting concerns in software development, explain aspect orientation as a technique that provides an additional modularization layer to handle crosscutting, introduce a formal framework for explaining aspect weaving and a composition model. As an outlook, this foundation may also rely on model level verification tools as well as in code and model weaving to remedy model (security) failures. In this way, we may also consider our work to be a contribution to model based development with aspects.

Contents

List of Figures

List of Tables

Listings

Chapter 1

Introduction

My main thesis is that science was created through the invention of critical discussion.[1]
-Karl R. Popper

Aspect Orientation (AO) is usually accepted as a novel software development paradigm. Indeed, aspect-oriented programming (AOP) was conceived in the early 90's at Xerox PARC. AO aims at solving some limitations of existing programming paradigms to achieve a better " separation of concerns" such as that many requirements do not decompose neatly into behavior centered in a single component[2] or modularization element. The result is that these kind of concerns are scattered across several modularization entities. A concern is according to the AOSD Ontology: "an interest, which pertains to the system's development, its operation or any other matters that are critical or otherwise important to one or more stakeholders." [77].

This way AO can be considered to provide one more dimension for modularization of software. The idea is that computer systems are better programmed by separately specifying the various concerns (understood as properties or areas of interest) of a system, describing their relationships, and then relying on mechanisms in the underlying aspect-oriented environment to *weave* i.e. compose them into a coherent program; see Example 2.2.

This description of AO is uncomplicated, though it conceals a number of problems.

---

[1]Meine Hauptthese ist, daß durch die Erfindung der kritischen Diskussion die Wissenschaft erzeugt wurde. Karl R. Popper in [63]

[2]we use the term component in a conceptual and logical sense, not as a programming or physical entity.

1.1 Problem Outline

There is however a number of problems related to the idea that computer systems are better programmed by adding a modularization layer with aspects and weaving them to the base system or model. On the one hand, current aspect-oriented techniques are independent of the crosscutting problem itself. In other words, the techniques may be used to solve problems we might have solved by means of conventional programming languages. Therefore, in some cases AO might not actually be needed, but the techniques themselves are used. We do not find clear guidelines of when should we actually apply aspect orientation and when we should not apply it. On the other hand, there is a wide range of interpretations of what aspect orientation means. Despite numerous techniques and methods around which all claim to support AO, there is still need for precise definitions.

Furthermore, aspect orientation techniques allow for modifying a software system by weaving in additional functionality. This brings a number of unsolved issues. For instance, the issue of code and model verification in aspect weaving is still open. Consider adding constructs on models and code that may interfere with the base functionality in undesired ways. We particularly refer to the problem of aspect interaction, which can also be regarded as a subproblem of feature interaction. These are the two main questions we deal with in this thesis. Namely, the concepts of aspect orientation and the problem of aspect interaction.

1.2 Main Achievements

In order to provide for clear definitions, we explore the source of aspects in view of tracing from requirements to the subsequent phases in software development, particularly from (informal) requirements to (formal) specifications. While reflecting upon the question " how can we define aspects (beyond the programming level)?" we came to the conclusion that crosscutting can be defined by the transformation from one abstraction level to another. We explore the source of crosscutting by means of a web store example that illustrates the translation of requirements to specifications. The model of the web store is built from the informal system requirements which we consider as belonging to the problem space, these requirements are translated into a system model based on classes of the object orientation and further specified in the lightweight formal language Alloy. We show that C^3 are specifications defined over several modules or classes. This way we provide aspects with meaning by defining them as entities that improve the base modularization paradigm i.e. entities that capture or implement C^3. We argue that there are two main kinds of crosscutting. This allows us to classify the aspects into two broad groups: functional and non functional. We can then focus our attention on functional aspects in Chap. 4 and show that these can be modeled by a special kind of refinement (namely communication refinement) which defines and explains "aspect weaving." We demonstrate (in Chap. 5) how a mathematical analysis of aspects allows not only to make some concepts more precise, like weaving and therefore aspect orientation but also to model particular problems, like aspect interaction. This analysis also establishes that aspect orientation (at least in the way we propose it) does indeed allow for better modularization. Equally important, our framework provides a view of AO in terms of layers of refinement which provides the foundation to define a composition model over communication channels between components (components considered in a logical way) and explain aspect interaction. This framework has as formal underpinning the FOCUS theory; see [11]. Derived from the above mathematical framework, we focus on the issue of aspect interaction in Chap. 5. It is an unresolved problem for which so far no conclusive results have been provided. Specifically, the possible interaction of different aspects as a result of the weaving process. Since there exist aspects that actually contradict each other (for example, the security aspects of anonymity and accountability), it is in general not clear whether two different aspects can be woven into a program without negative interference. Equally important, establishing this for a given set of aspects and a given program is a non trivial challenge which requires having a formal foundation to analyze the system and the aspect interaction. Specifically, the problem of aspect composition involves analyzing not only their execution ordering but is rather a problem of semantic analysis of the aspects that are woven on a given base element (such as an object's method, attribute or the interface of a component).

To explore the problem of aspect interaction we draw inspiration from one of the main research lines in the field, the Composition Filters (CF) approach [7]. The reason is that this approach allows for a particularly insertion of aspects over the communication channels between objects or logical components. We recall that every system i.e. subsystem (in the sense of logical system entities) interaction can be reduced to the receiving and sending of messages. Making the communication channels a first-class entity in the aspect interaction analysis helps reasoning over a number of problems in aspect-oriented software development. We chose to explore aspect interaction in this work at the hand of two security aspects from industry interesting enough for a formal mathematical analysis. In addition, we recall that aspect interaction happens actually at specific points, this is why we show our work on a few well defined points of execution.

Furthermore, consider the definition of *clear* and *black box aspect-oriented programming* (AOP) from [22] in terms of quantifications over the internal structure of components (clear box) or over the public interface of components (black box). In both cases, we have communication channels, either external (at the interface level) or internal (in the decomposition of the subsystem). We address the specific situation where one can weave aspects into the interfaces among components, also known as *black box aspect orientation*. As mentioned above, it is related to a line of research in AO that adds aspect behavior through input and output filters superimposed on sets of objects.

Altogether, we identify the source of *crosscutting concerns* (C^3) in requirements, explain *aspect orientation* as a technique that provides an additional modularization layer to handle C^3 together with composition or weaving mechanisms, propose a formal framework for explaining aspect weaving, and a composition model that allows to analyze composition of multiple functional aspects at particular execution points.

1.3 Our Approach and Related Work

Crosscutting has been analyzed and previously defined in the work of Kiczales et. al [51]. The work of Kiczales in aspect orientation which is one of the cornerstones in AOP. However, it is rather related to programming and we prefer to relate to a conceptual framework that helps to explain the transformation steps from requirements to the other development phases, and actually provide meaning to such concepts; see Table 1.1. We also propose to discriminate C^3 from aspects, the former being the reason for the existence of the latter. An exploration of the meaning and use of aspect orientation has not been published so far at least in an integral manner as here. By this we mean, examining its (essential) nature and showing when and how it should be used.

With respect to aspect interaction, most approaches address the aspect interaction problem at the programming and language level such as the work of Durr et al in [18]. However, the particular issue of semantic analysis of compositions at the (more abstract) modeling level in the style of composition filters is still an open question, and to the best of our knowledge has not been formally achieved by combining CF and the FOCUS formal theory that provides a syntax and semantics to model (logical) components; see [11]. Related work on aspect interaction can be found in [6], however the authors restrict themselves to a global overview of the problem and their results are not comparable to ours. Nagy et al [59] explored aspect interaction and proposed a set of requirements for composing aspects at so-called *shared join points*, although the work is rather directed to a syntactical level. Their later work [32] analyzes aspect composition by graph transformation which is a different technique as the one we used here.

Although a formal foundation for aspects exists for a subset of Java and Aspect/J [80], no formal foundation has been published so far for an approach as composition filters and aspect weaving at the software architecture level. This is one of our contributions.

Table 1.1: Three possibilities of Aspect

Type I	Cross-cutting concern	requirements level
Type II	Base model plus optional element(s)	architectural level
Type III	Base element(s) and additional construct(s)	programming level

1.4 Thesis Structure

In short, we provide a conceptual and later a mathematical analysis of aspect orientation, define a model for aspect composition and illustrate its use at the hand of two particular security aspects. We also show how this framework allows to analyze the conditions under which aspect composition respects desired functional properties (according to the individual specification of each aspect before composition) at a semantical level.

This work is structured as follows:

- □ Chap. 2 starts with an historical view on aspect orientation, then gives an overview of selected aspect-oriented programming languages. This provides the background for our framework explaining aspect orientation in more general terms.

- □ Chap. 3 allows us to consider the key elements of AO in programming level approaches, and analyze these intuitions at a more (architectural) level. Following a research on the nature of aspect orientation we differentiate C^3, aspect, and provide a definition for both. The definition of the former together with a classification is provided.

- □ Chap. 4 provides the formal underpinnings we use to explain aspect weaving and build our formal composition filters model in the next chapter.

- □ Chap. 5 provides a formal model for aspect composition together with an application of the framework at the case of a non trivial problem in the field, namely the problems derived of weaving multiple aspects. Primarily, we focus on the issue of aspect interaction which can be related to a similar problematic: feature interaction. The chapter draws a parallel between model weaving and code transformation.

- □ We close this work with some conclusions in Chap. 6 together with an outline of future lines of research.

Chapter 2

Aspect-Oriented Programming

Chi va piano, va sano e va lontano.[1]
-vox populi

In this chapter, we recall some aspect-oriented programming (AOP) languages together with an overview of related approaches. The goal is to provide a basic understanding on the subject. Also, to illustrate that the techniques behind it are, to a great extent, language additional constructs that allow for code transformation, either dynamically or statically. We may consider AOP a special form of code transformation, specifically code transformation through *weaving*. Subsequently, we may direct our attention to weaving from a systems architecture perspective and draw a parallel between weaving and model transformation[2] or at a more mathematical level to refinement as introduced in Chap. 4.

2.1 Historical Perspective

A historical overview indicates that AO is actually not a new idea[3] despite being considered a novel paradigm. Again, by approaching it from a mathematical perspective, we discovered that the techniques behind it have been in use for quite some time.

According to [70, 79], the techniques preceding it are: layers, collaborations and subjects. The motivation for these concepts is in most cases to support the principle of *Separation of Concerns* (SoC). Following from this principle the work in AO "is based on the belief that programming languages based on any

[1] Who slowly goes, makes sure and far

[2] In Chap. 5

[3] See for instance the article by I. Jacobson [42], showing that his work in the 1970's contained concepts today considered *aspect oriented*, although under different names.

single abstraction framework - procedures, constraints, whatever - are ultimately inadequate for many complex systems." [51]

In the following we shortly survey the related concepts.

Collaborations and Layers are defined by a set of roles and a communication protocol between the classes participating in the collaboration [17]. This means that the classes pertaining to a set of cooperating classes communicate and interact with each other based on the set of roles and a communication protocol defined by a unit that is called "collaboration." The goal is to represent an application as a composition of collaborations that can also be represented as layers. Smaragdakis and Batory [70] explain collaborations and layers as a view of an object-oriented design from the perspective of a single feature or "concern." Each collaboration or layer would capture distinct functional aspects of an application. According to the authors this is a kind of refinement that extends a program's functionality or adds implementation details to it.

Subjects are another alternative for enhancing the composition mechanisms in object orientation[4]. A *subject* is a collection of classes or class fragments related by inheritance, aggregation or association. A *subject* is a partial or a complete object model that corresponds to a perspective[5] on the objects to be grouped as such. A perspective can be the result of the usage context of an object although this may not be the only reason. A subject may also emerge from the points of view of different development teams [17].

Aspect-oriented programming (AOP) can be defined as a set of techniques supporting code transformation by means of additional language constructs. These constructs allow the diverting of the execution flow of given code through a mechanism (*weaver*)[6] that combines some *base* code (i.e. the originally existing code) with *aspect* code (i.e. the additional code) as represented in Fig. 2.1. Therefore, the aspect languages are additional constructs and some mechanism(s) that support code transformation in ways not available before. Consider this for example, realizing interception of messages among objects (as illustrated in Example 2.2.4) without these techniques. This is, for us, the key idea behind AOP. We explore it at the programming level[7] in this chapter. In the following chapter we examine the concepts of AO from a broader systems development perspective, in other

[4]The composition mechanisms in object orientation are: inheritance and aggregation.

[5]Again we find here one the main motivations behind aspect orientation. The name aspect also hints to this meaning.

[6]We relate weaving to a transformation function in Sect. 4.2.3

[7]What we call a *bottom up* approach.

words, from the perspective of how a system is subdivided in subsystems and how these relate to one another; namely an architectural perspective. This means, we identify the source of crosscutting in requirements and show the way C^3 surface in a given design; *aspects* are then defined as the entity that allows to encapsulate C^3.

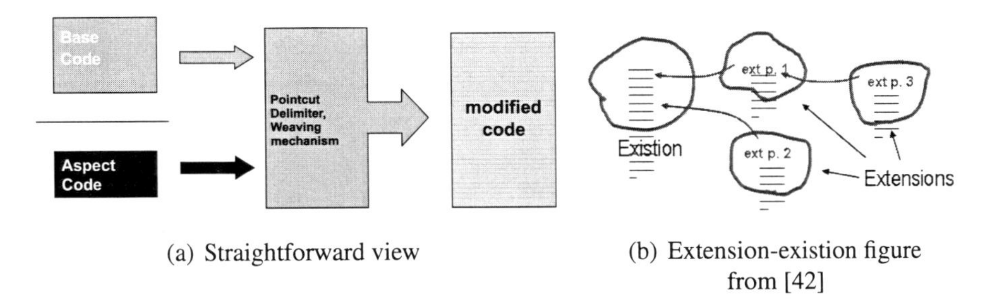

(a) Straightforward view

(b) Extension-existion figure from [42]

Figure 2.1: Weaving notion

Pointcut delimiter model	Weaving	Language
Group I. Join Point + Advice	static / dynamic	ASPECTJ
Group II. Message interception via filters	static / dynamic	COMPOSE*
Group III. Multidimensional SoC	static	HYPER/J
Group IV. Traversals	d.K.	DEMETERJ

Table 2.1: Selected aspect-oriented programming languages

We can now describe some aspect-oriented languages grouped according to the way the elements of the base model are selected, in other words according to their *pointcut delimiter model* (see Table 2.1). We considered these to be the most representative of AOP following the approaches mentioned in [20].

2.2 Group I: Join Point and Advice

This group extends a base language[8] with the goal of adding some additional behavior to the one originally provided. This may be achieved at run time or by com-

[8]For instance, Java

pilation. The former case is called dynamic and the latter static. The additional constructs offered by this language group are: *Advice*, *Introduction*, *Join Point* and *Pointcut*. We explore these examples in Example 2.2 and Example 2.2.4.

Introductions are also known as intertype declarations. These are declarations about a program's structure that provide a mechanism for adding new static as well as behavioral elements in the class structure. According to this, an aspect can be used "to add new methods to a class, or to declare that a class extends to a super class" [28].

Join Points (JP) are special well defined points in the execution flow of the program. We may define them as syntactical elements that allow to discriminate or select code elements like: method calls, object creation, exception execution, and attribute value assignment. As such, JPs have no behavior, since they are only an element that allows for selecting the elements at which additional behavior will be added (or weaved).

Pointcuts help to refer to some set of join points whose execution needs to be modified. A pointcut characterizes one or more join points along with its contextual values [53].

Advice is the code that is executed when pointcuts are reached. It is the actual implementation of an aspect. The developer must determine when the advice is to be executed. There are three execution ordering possibilities with respect to the base code: *before*, *after* and *around*.

The above language constructs shape what we may call the *Join Point (JP) and Advice Model*. It is precisely the concept that supports indicating where some additional behavior should fit in the given code. This model is related to the programming level but at the same time it has also been used to explain aspects at the modeling level. It is worth noting that most of the work on aspect analysis and interaction focuses on these constructs.

Example 2.2. Undo Assume we need a method that allows classes (i.e., objects) in an existing code to restore their content after any set operation on their data. We suggest performing this by inserting an `undo()` method on selected classes. The `undo()` method allows then for a way to reverse the changes carried out in the objects (in an object oriented program). This method might imply modifications of the selected classes in behavior and structure, in case we choose not to implement it as an aspect. It modifies a class structure using an aspect and interface. The initial program structure is preserved and the aspects are woven in what we may describe as a *non-intrusive* manner, in other words, the base code remains unchanged. The intent is to create a framework for preserving an object's history, thus allowing for restoring, if required, an object's previous contents after a given method has been executed. We assume that the object's encapsulation is not violated, in other words, we may only alter an object's content through its set methods. Still, we need to guarantee that only one method may actually perform a "set" operation on the concerned object. The `undo()` method is expected to keep track of methods that change the contents of an object.

We consider Example 2.2 a crosscutting problem that may exemplify similar cases. We designed this small example because it is also somehow more interesting than the usual *logging* aspect we find so often in the literature. Our solution is outlined in Fig. 2.2. In brief, whenever a `set` method of one of the selected objects is called, this call is intercepted and a clone of the object is created. After this, the intercepted method is performed (see the label «proceed» in Fig. 2.2). After the originally called method is performed we have the object itself (now modified) and a clone representing its content previous to the changes. At this point, we may decide whether to leave the changes or to discard them. Relating it to Fig. 2.1, the **base code** is the set of previously given classes e.g. `Circle` (Listing 2.1), the **aspect code** is the one in Listing 2.4 in which we also find the pointcut delimiter (line 5 of the listing). The weaving mechanism is the ASPECTJ compiler. We will get back to this concepts in the coming subsections.

Such crosscutting concerns might not be specific to a particular class yet it is implemented for Java code (given the language we wish to illustrate). We outline an example for the sole purpose of explaining ASPECTJ, which is by the way, considered the *de facto* standard in AOP. The base classes are `Circle` and `Square` (see Listing B.2 in Appendix B).

ASPECTJ This aspect-oriented programming language is an extension of the Java object-oriented language. This extension is accomplished through the concepts mentioned above: *introductions*, *pointcuts* and *advices*. It is conceived as a general purpose aspect language.

Its use belongs mainly to two broad categories: Development Aspects and Production Aspects [50]. The former, when dealing with developing aspects means that the language can be used to facilitate tasks such as: debugging, testing and performance tuning of applications. The latter, when dealing with production aspects, can be used for implementing functionality intended to be included in applications.

In laymen terms, the particular constructs of ASPECTJ are usually applied once the base classes have been developed. Then, aspects are written and woven on it.

Listing 2.1: Class Circle

```
class Circle{                                                    1
 private int rad;
 private int x=0;                                                3
 private int y=0;

...
 public void moveCircle(int newposx, int newposy) {              7
  x = newposx;
  y = newposy;                                                   9
 }
...                                                              11
}
```

2.2.1 Introductions

Introductions, also known as *intertype declarations*, are declarations about a program's structure that provide a mechanism for adding new static as well as behavioral elements within the class structure. We may add new methods to a class, declare that a class extends to a super class or that it implements an interface that was not originally considered or defined. Some also refer to these as intertype declarations which affect the static hierarchy of programs. An example of an introduction that modifies the base classes in Example 2.2 is illustrated in Listing 2.2, specifically lines 2 and 3. The line `declare parents` relates to the base class and the interface added to it by declaration `implements`.

The `clone()` method in the main program is used in Listing 2.3.

Introductions allow the programmer to modify classes by adding fields, methods, and even changing the hierarchical structure by a sort of inheritance. The elements that can be declared by using intertype declarations are: Fields, constructors and concrete or abstract methods.

Now, let us explore some constructs in ASPECTJ that can modify code execution[9]: `Pointcuts` (as sets of `Join Points`) and `Advices`.

2.2.2 Join Points

Join Points (JP) are special well defined points in the execution flow of the program. These include method calls, object creation, exception execution and at-

[9] as already mentioned, there are two dimensions at which we may have aspect weaving. One is static, when the code is compiled with some weaving mechanism and then executed. The other is dynamic, when the code is interpreted in run-time and the running environment provides a way to dynamically select the join points and execute the advices accordingly

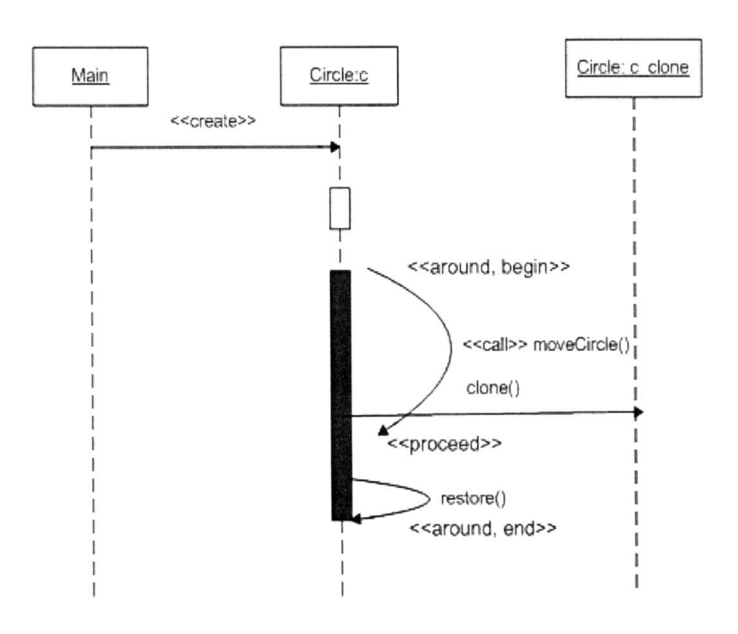

Figure 2.2: Sequence of the **undo**() method

tribute value assignment and others. Each one has a special syntax. For example, let us examine Listing 2.4. This example contains the three elements that we will explain in the following: `join point`, `pointcut` and `advice`. Let us now focus on *join points*. Taken apart from the rest of the code, in line 5 we define the syntactical element related to the execution point at which the aspect code, later referred to as "advice", will be bounded.

2.2.3 Pointcuts

A pointcut provides an identifier for a set of JP's that can actually be referred to, and modified, by several advices. Pointcuts help to refer to some sets of join points along with its contextual values where the execution flow will be modified. For instance, in line 5 of Listing 2.4 we select every call to method `moveCircle` in class `Circle` and give it to the identifier `tracePointsCircle`. We may use this identifier to relate the pointcut to the code that will be executed when it is reached. That is to state in our example the code specified in lines 8 through 10.

This brings us to the question of when and how the attached code be performed.

Listing 2.2: Introductions in ASPECTJ. Cloning interface added to class Circle

```
public aspect addCloneMethod {
   declare parents: Circle implements Cloneable;
   declare parents: Square implements Cloneable;
   public Object Circle.clone() {
   try {
    return super.clone();
   }
   catch (CloneNotSupportedException e) {
    throw new InternalError(e.toString());
   }
  }
   public Object Square.clone() {
   try {
    return super.clone();
   }
   catch (CloneNotSupportedException e) {
    throw new InternalError(e.toString());
   }
  }
}
```

2.2.4 Advice

These represent the actual implementation of the aspect. As already mentioned, there are three significant modifiers that determine the execution order of the *aspectual* code (i.e., advice), namely *before*, *after* and *around* the selected pointcut. We explain these in Example 2.2.4.

Example 2.2.4. Bank payment through web Consider the case of a client paying through a web browser. This is implemented by three classes: *customer, web* and *bank*. The sequence diagram for the base case is shown in Fig. 2.3. The

Listing 2.3: Using the added interface in ASPECTJ

```
public class cloneDemo {
 public static void main(String args[])
  throws CloneNotSupportedException {
   Circle obj1 = null;
   Circle obj2 = new Circle(15);
   obj1 = (Circle)obj2.clone();
   System.out.println(obj1.getradius());
 }
}
```

Listing 2.4: Sample code listing in ASPECTJ

```
package clonning;                                                        1

public aspect addCloneMethod {                                           3

pointcut tracePointsCircle(Circle c) : target(c)  && call(* ↩           5
    moveCircle(..));

void around (Circle c) : tracePointsCircle(c) {                          7
  println("Intercepted message MOVECIRCLE: " + ↩
     thisJoinPointStaticPart.getSignature().getName());
  println(" in class: " + thisJoinPointStaticPart. ↩                    9
     getSignature().getDeclaringType().getName());
  Circle backupCircle = (Circle)c.clone();
  proceed(c);                                                            11
}
```

customer pays *(credit card)*, inputs card data, then the browser (*web*) executes method *payment()* at the bank, the bank responds with *ok*, and finally *web* sends a confirmation to the customer. This is straightforward; we did not consider any other possible scenarios. For instance, credit card reject from the bank, security issues, not enough money in the customer's account, etc. We propose to perform the following three modifications on the base case:

1. An authentication protocol before the payment.

2. Notify a third party (say the warehouse) that the payment has taken place.

3. Do not pay directly at the bank, rather through a third-party system (take for instance "paypal").

In the following, we present the ASPECTJ constructs that determine the ordering of advice execution with respect to the triggering action, i.e., reaching the pointcut.

before In order to perform an additional protocol, say authentication, before actually sending the credit card information to the bank, we may select the payment method `placePaymentOrder` by its signature with the code in Listing 2.5. This means, the signature of the method is the class to which it belongs, its name, and its parameters. This is indicated after the keyword call in the pointcut. The modifier `before` in line 3 defines the moment at which the method is intercepted and the authentication should take place. This is illustrated in Fig. 2.4.

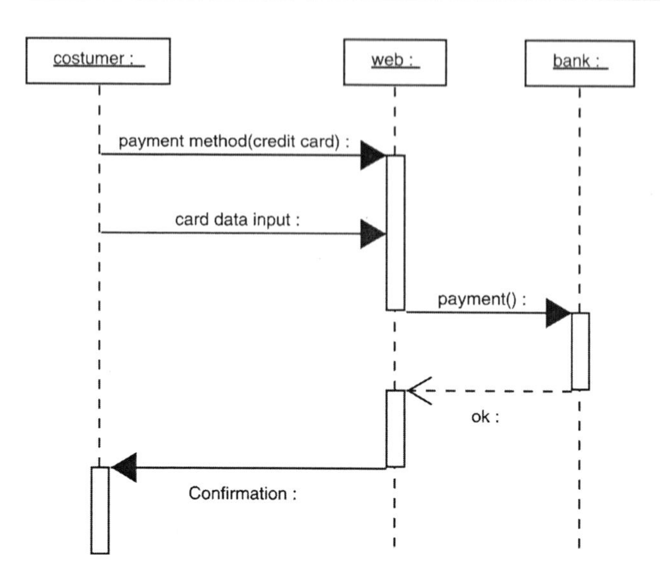

Figure 2.3: Bank payment example. Base case. Sequence diagram with no advice

Listing 2.5: Bank payment example. Before advice for payment(), short version

```
public pointcut tracePayment(Bank bank): call(int Bank. ↩
   placePaymentOrder(..)) && target(bank);

 before (Bank b) : tracePayment(b)
  ...
```

after Suppose now that we want to add a method that notifies the warehouse when the payment has succeeded; see Fig. 2.5. To achieve this, we may write an aspect, for example Listing 2.6, where in line 3 we would provide the corresponding advice (sending some message to the warehouse). The keyword `after` and the pointcut indicate the ASPECTJ compiler where such method would be *woven*.

around, last but not least, assume we decided to override the current payment method and substitute it with another one, namely *anotherPayment* in Fig. 2.6.

Listing 2.6: Bank payment example. After advice for payment(), short version

```
public pointcut tracePayment(Bank bank): call(int Bank. ↩
   placePaymentOrder(..)) && target(bank);

 after (Bank b) : tracePayment(b) ...
```

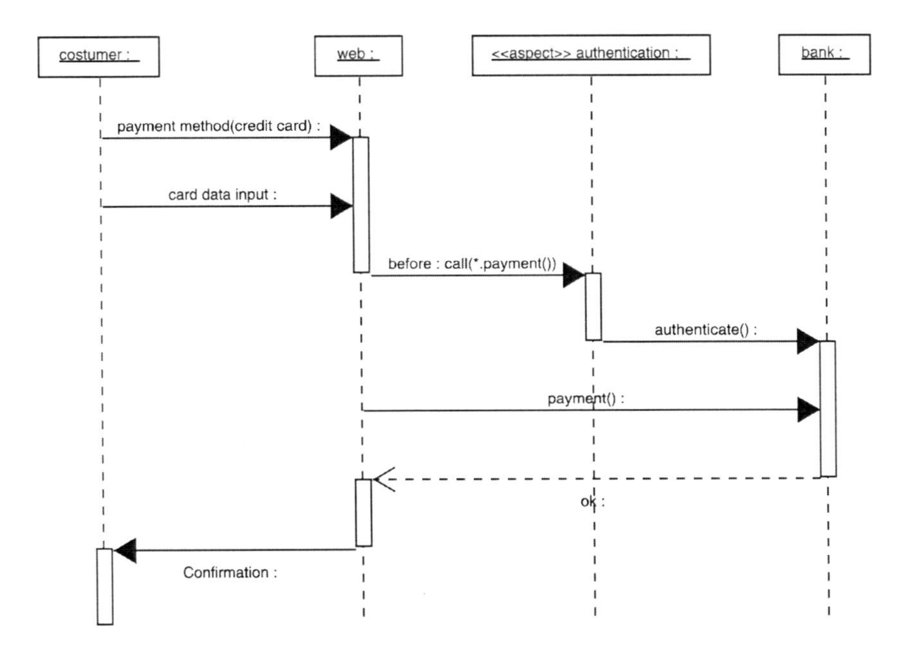

Figure 2.4: Bank payment example. Sequence diagram with *before* advice

This aspect prevents the original method from being called. The pointcut is the same as indicated in the listings before but the keyword around (in Listing 2.7) indicates that this advice executes instead of the original method. Please note the type delimiter "int" before the keyword in this case, here we provide a return type for the aspect.

Furthermore, in Example 2.2 we mentioned the problem of keeping a record of changes in the base classes. An around advice is also helpful in that case to copy the original content of the classes, execute the originally called method and then by the keyword `proceed()` return to the point where the program flow was deviated and continue execution of the base code. This is a variation of the around advice in Listing 2.4. This variation allows to execute the method that was " circumvented". In the Example 2.2.4 the circumvented method is `payment()`.

Listing 2.7: Bank payment example. Around advice for payment(), short version

```
...
int around (Bank b) : tracePayment(b){...
```

One issue with this language is, as expected, that the types of methods and parameters has to be respected and considered at the time of writing the aspects in order to conform to the base program. ASPECTJ is no magic bullet that may

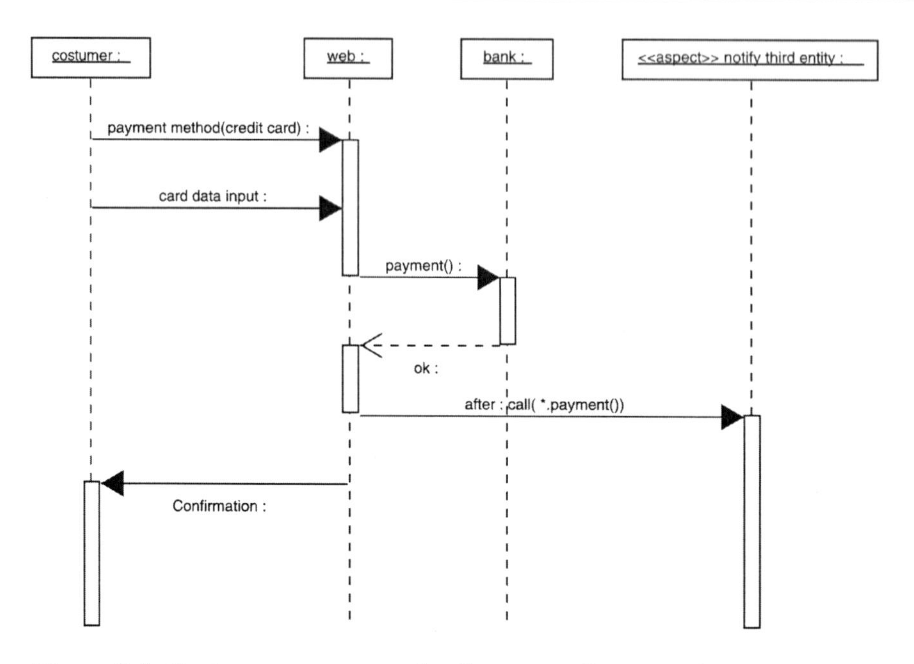

Figure 2.5: Bank payment example. Sequence diagram with *after* advice

overcome such limitations.

In Listing 2.4, the moment of execution of the advice is specified by the key word `around` in the pointcut specification. The advice itself is defined as Java code inside the brackets within the same listing.

There are other characteristics of this language, although the goal of this section was to explain the three constructs that make ASPECTJ different from its base language (Java), once again these constructs are:

- Join Point, Pointcut,
- Advice,
- Introduction

The *Join Point+Advice Model* has been also extended to other languages such as Smalltalk, C++, for instance. It plays an important role in defining the terminology used in the AO community (as can be inferred by reading the ontology in [77]) as well as in most research within the field. The vocabulary in the AOP community is heavily influenced by this model, namely the ASPECTJ way of dealing with aspects. We believe most of the terms to be a good starting point though reducing the problem of aspects to the sole approach of the JP model and terminology might leave aside approaches as "Multiple Views" and may lead to believe

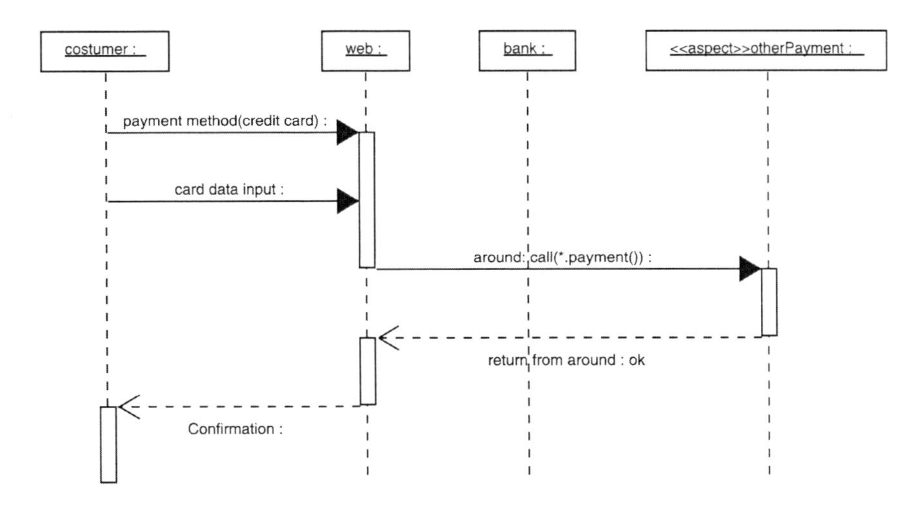

Figure 2.6: Bank payment example. Sequence diagram with *around* advice

that the sole use of JPs and Advices makes a design or implementation *aspect-oriented.*

2.3 Group II: Message Interception

In this section, we describe a composition (or program modification) language that is based upon message interception. Concern composition and modularization is accomplished by extending the object-oriented model through message interception via *filters*. These filters seize messages sent among objects. The messages are selected on logical conditions quantified over their signature, the signature of the sender or receiver.

The reason to extend the traditional object model through message interception is because in an object-oriented program all observable, i.e., externally visible behavior of an object, is revealed in the messages in that it sends or receives. Message passing is the basic means for creating executions in a software system as demonstrated by [3].

We recall the language constructs of *Composition Filters* from its manual [74] and the work of Bergmans et. al [7]. We explain this language at the hand of some examples.

Composition Filters (CF) are motivated by the need to express any kind of message coordination in the conventional object model. CF's extend the object model with a number of message filters.

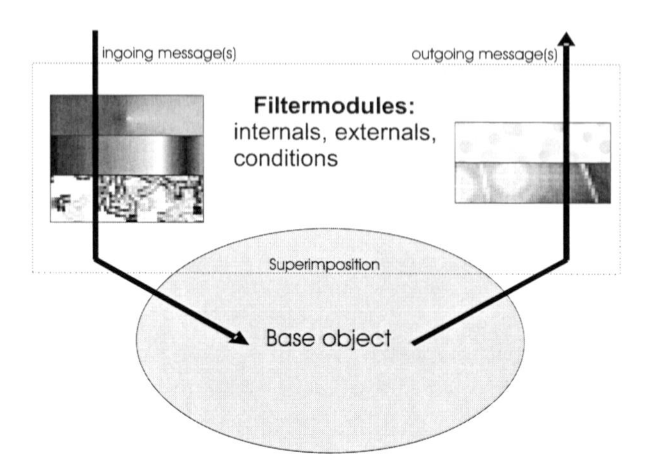

Figure 2.7: Representation of a filter module [74]

In the CF model, the object is composed of an interface and a base component. The base component is the object itself and the interface is added in order to trap incoming or outgoing messages coming to or from the object, this is represented in Fig. 2.7. This allows the modification or redirection of messages to external or internal objects according to predefined "filters". This technique was originally motivated by the need to support the implementation of synchronization and real-time constraints, atomic transactions, precondition style error checking and other aspects in a well localized way.

Following the argument that aspect languages are constituted namely by constructs extending a base one, we explain CF at the hand of its message interception idioms. The constructs of this language are mainly:

- **concern** with filter modules, superimposition, and implementation parts,
- **filtermodule** with internal and external objects, conditions, and the filters;
- **superimposition** with class selector(s) and their relation to the filter module.

2.3.1 Concern

It is the main building block that extends a base application. It is the place where filter modules are specified and also where these are related to specific points

Listing 2.8: Filter module and superimposition in COMPOSE*. Example 2.2.4

```
concern PaymentSecurity in dress4Less
{                                                                      2
 filtermodule securePayment
 {                                                                     4
 internals
  authentication : dress4Less.SecurityProtocols. ↩
      authenticate;                                                    6
  encryption : dress4Less.SecurityProtocols.secureChannel;
 conditions                                                            8
  authenticated : dress4Less.SecurityProtocols.authenticate ↩
      .clientAuthenticated();
 inputfilters                                                          10
  authenticate_filter : Dispatch = {True => [*. ↩
      placePaymentOrder]authentication.paytobankAuth,[*. ↩
      choosePaymentMethod]authentication.choosePaymentAuth};
  encrypt_filter : Dispatch = {True => [*.sendAuthenticated ↩       12
      ]encryption.encChannel}
 }

 superimposition
 {                                                                     16
  selectors
   payment = { *=dress4Less.Bank, *=dress4Less.Web, *= ↩             18
      dress4Less.SecurityProtocols.authenticate};
  filtermodules
   payment <- securePayment;                                           20
 }
}                                                                      22
```

(in the superimposition part) of the base program which can be seen as the base concern. It has three parts, all are namley optional, the filter module, the superimposition and the implementation.

We may illustrate this with the help of Fig. 2.7. Thus visualizing the concern as the whole of this figure. There we find the base object which might be any usual object in object orientation in the part below. Over it we find a rectangle containing the so called `Filtermodule` with a left hand and a right hand side. These are the filters (input and output, respectively) represented as the rectangles through which the messages (please notice the arrows) flow to the base object. The language construct indicating which sets of filters are applied over a selected base object is the `superimposition`.

2.3.2 Filter Module

Is a collection of (zero or more) filters, where the filters are the entities selecting and/or modifying a message that is directed toward the base object or that is sent from it. Filters define conditions so that the messages are either selected and some action is associated to the message, or are left to pass through without any action being take.

In Listing 2.8 between lines 3 and 7 we have an example of a filter module. There are two internal objects: `authenticate` and `secureChannel`. These are the security protocols we add to the base program and are written in a DotNet language, we show the dummy methods in Appendix E.1 Listing E.1 and Listing E.2, respectively.

From lines 10 through 12 we define the filters and mark them as input filters. This means, they act on messages directed to the object(s) specified in the second part of Listing 2.8. These filters are: `authenticate_filter` and `encrypt_filter`. Let us have a closer look at the second one in Listing 2.9.

Listing 2.9: Part of concern payment: Encryption filter

```
encrypt_filter : Dispatch = {True => [*.sendAuthenticated] ↩
    encryption.encChannel
```

There are four types of filters: *Dispatch*, *Send*, *Error* and *Meta*.

Dispatch filters substitute the target and selector (`*.sendAuthenticated` in Listing 2.9) with the target and selector of the substitution part (`encryption.encChannel`), in case the message is accepted. In this example, the selected method is substituted by `encChannel` in Listing E.2. We may specify some conditions (in a method returning a boolean value) instead of `True` in this examples, in case necessary.

Send filters are "the dispatch filter for the output filters." We did not try output filters in our example, because of implementation issues in the different versions of COMPOSE*.

Error filters raise exceptions when there is a rejection.

Meta filters reify the current message and add it as an argument of a new message when the filter is matched. We did not try these kinds of filter for the above mentioned reasons.

2.3.3 Superimposition

This is the construct where we tell COMPOSE* to bind filters to other program elements. In the figurative sense, the superimposition actually puts the bigger rectangle (containing the filters) from Fig. 2.7 above the base object(s). This happens, as stated in the manual, during initialization of the application.

Examine, for instance, lines 18 through 20 in Listing 2.8. We first give a name, in this case `payment` to a collection of classes and their methods (`*=dress 4Less .Bank`). The latter means that we select every method in class `Bank` of package `dress4Less`, we then indicate that the filter module `securePayment` will be "operating over" the collection of methods delimited by the selector `payment`. It is possible to relate other filter modules to other selectors.

Our intention in this lines is to provide for an overview of the main constructs in COMPOSE*, not an exhausting revision. That is what manuals are actually for.

2.4 Group III: Multidimensional Separation of Concerns

The motivation behind this approach is the need to manage several kinds of concerns "uniformly". As compared to actual technologies that make a kind of concern (say also modularization criterion) the prevailing one, e.g. classes, components, "services", etc. This implies that other criteria are subordinated. Multidimensional Separation of Concerns (MDSoC) defines a modularization level that is built over the class abstraction taking it as a basic building layer. More than that, it has the goal of identifying, encapsulating and manipulating only those parts of software that are relevant to a particular concept, goal or purpose. This concept, goal or purpose is then handled as an "orthogonal" layer that defines a grouping of classes. Within this grouping it adds the (functional) characteristics relevant to the concern in question. We recall that in the case of object orientation the main concern, in sense of being the first modularization concept, is the data concern i.e. classes. The promoters of this approach consider that there are other kinds of concerns such as: features, aspects, roles and viewpoints. MDSoC as in [62] supports the following characteristics.

- □ Multiple, arbitrary dimensions of concern.
- □ Separation along these dimensions simultaneously.
- □ The ability to handle new concerns, and new dimensions of concern, dynamically, as they arise throughout the software life cycle.

□ Overlapping and interacting concerns.

As already mentioned, in MDSoC a kind of concern is considered a dimension of concern. MDSoC therefore decomposes software according to several dimensions of concern or features. Among the goals, assumed by the initiators of this programming approach, is the simultaneous and incremental decomposition of concerns as well as being able to represent and manage overlapping and interacting concerns. Say for instance, in the web payment example, Example 2.2.4, the authentication protocol might be set inside of a concern (security) and handled as an additional layer over the base set of classes.

Moreover, the problem this approach tries to solve is the so called *tyranny of the dominant decomposition*. This is to say that in the OOP the class is the dominant concern and it hinders programmers from neatly separating other kinds of concerns like features or aspects. We further elaborate on this insight in Sect. 3.3.

HYPER/J is the implementation of multidimensional separation of concerns. The main additional constructs of this language are *hyperspaces*, *concerns*, and the setting together of both is indicated in a *hypermodule*. Hyperspaces support the goals of MDSoC by allowing identification of any concerns of importance, encapsulation of those concerns and management of relationships among them together with their integration. It is implemented by the language HYPER/J which supports the concept of hyperspaces in Java. It may extend, integrate and change the original modularization of previously existing programs without having to modify, or even have, source code.

We explore it with the example provided in the manual [73], the Software Engineering Environment (SEE). The reason is the example is designed in a way that each concern dimension is "declaratively complete." This means, in case an operation is defined in some concern dimension and this dimension is not selected for composition, then the resulting classes are not executable, as we would have naturally expected. The example from the manual is designed to be easily refactored.

The constructs of this language are: *hyperspace*, *concerns* and *hypermodules*. Each is defined in a file (though we may define them altogether in just one text file) that is passed as a parameter to the Java compiler together with the tool itself (com.ibm.hyperj.hyperj, see below)

As we may see in Listing 2.10, compiling a software in HYPER/J is a batch kind of work, though there will be a graphical user interface in the coming future for this tool.

Listing 2.10: HYPER/J demo compilation

```
java -cp %JAVA_DIR%\jre\lib\rt.jar;%JAVA_DIR%\lib\tools.jar ↩
    ;%HYPERJ_DIR%;%HYPERJ_DIR%\bin\hyperj.jar com.ibm.hyperj ↩
    .hyperj
-hyperspace %HYPERJ_DIR%\ObjDimDress.hs
-concerns %HYPERJ_DIR%\dress4Less\concerns.cm
-hypermodules %HYPERJ_DIR%\dress4Less\Order.hm -verbose
```

2.4.1 Hyperspace

According to the definition in [73] "A hyperspace is a concern space specially structured to support our approach to multidimensional separation of concerns." Now, a concern space encompasses all units in some body of software. Finally, a hyperspace is an identifier for the units we desire to handle. In the particular example, all the classes in the package `demo.ObjectDimension` (Listing 2.11).

Once we have established the collection of classes on which the composition will be carried on, we associate methods to "Features," where the features are the identifier for the groups of methods we desire to handle as such also called groups. The reasons may be: for evolving the software later, for providing different features according to some product line, etc.

2.4.2 Concern

It is the name given to the file relating methods in the classes, previously indicated in the hyperspace, to a kind of label (called "feature") that allows us to select the set of features for a given software product.

Take for instance, Listing 2.12, here we define three features or characteristics in the SEE: Check, Display, and Eval. Lines 3 and 7 indicate that operations `check` and `check_process` are associated to the feature `Check`. All three sets are designed in such a way that they may be compiled independently, in the sense that if we select only one of these features we may still have a running system (or we have to select the features that do constitute a workable system). The example is a simple equation or calculation solver. `Check` encompasses the operations that validate whether an expression is correctly written according to some predefined rules; `Display` contains the methods that show the result of the

Listing 2.11: Hyperspace definition (HYPER/J demo)

```
hyperspace DemoHyperspace
 composable class demo.ObjectDimension.*;                2
```

Listing 2.12: Concern definition (HYPER/J demo)

```
package demo.ObjectDimension : Feature.Kernel

operation check : Feature.Check
operation display : Feature.Display
operation eval : Feature.Eval

operation check_process : Feature.Check
operation display_process : Feature.Display
operation eval_process : Feature.Eval

operation process : Feature.None
```

operations; and `Eval` the operations that actually execute the calculation.

2.4.3 Hypermodule

Finally, a hypermodule defines the selection of features which comprise the desired system. Say, we may indicate HYPER/J to mix only Check and Display, not Eval, as we do in Listing 2.13.

According to Listing 2.13 the resulting system (.class files), will not calculate the results of expressions, but the expressions will be correctly written: Check-"ed", and Display-"ed."

2.5 Group IV. Traversals. Adaptive Programming

We briefly explore this other group, given that *adaptive programming* is also considered to be related to *aspect orientation*. We explore the meaning of adaptability and adaptive program. The concepts of this subsection are recalled from [55].

> A program is called adaptive if it adapts its behavior according to its context

There are several degrees of adaptability that depend on the context or frame of reference taken. Lieberherr et al [54] considers the following artifacts as context: data structures, class structures, data models, inputs to a program, run-time environment, software architecture, computational resources, for instance. This means, the goal of AP is to produce programs that adjust its behavior in relation to one or more of these parameters.

Listing 2.13: Hypermodule definition (HYPER/J demo)

```
hypermodule DemoSEE
 hyperslices:
  Feature.Kernel,
  Feature.Check,
  Feature.Display;
   relationships:
   mergeByName;

   equate operation Feature.Kernel.process,
    Feature.Check.check_process,
    Feature.Display.display_process;
end hypermodule;
```

Say for instance, an adaptive program with respect to a data model is one that is written for a generic data model and is restricted by a set of structural constraints [54].

An adaptive program consists of three parts: a graph C, an initial behavior specification expressed in terms of C, and behavior enhancements also in terms of C. Where graph C and the initial behavior correspond to components (we suppose the authors refer to logical components), and the behavioral enhancements are the aspects that add behavior.

How does AP achieve adaptability? By means of "collaborating views that are loosely coupled." We understand this statement as follows: A view is the building block in AP, collaborating views are the ones containing partial information between them, and louse coupling means that the dependencies between views are minimized. It seems in our opinion to be an *oxymoron.*

AP is implemented by a tool called DEMETERJ and its concepts have been expressed in the following terms: *traversals*, *class graphs* and *visitor objects*. Let us explore these concepts and try to understand the relation of AP to AOP.

An adaptive method in DEMETERJ is defined by parameterizing a program by a class graph. Class graph predicates specify traversals through objects of the class graph and define traversal enhancements. A *class graph* object is a graph whose nodes are classes and whose edges are "is a" and "has a" relationship between classes. This object has two methods: `traverse` and `fetch`. Method `traverse` is used to interpret traversal specifications, put in other words, it starts referring to a given object and crosses or "traverses" specified paths. We understand it as a pointer to the first object in the graph and a predicate (the traversal specification) selecting the next objects in the graph, i.e., the traversal specification is a relation as mentioned in the first part of this paragraph. Method `fetch` retrieves one object along one path in a Java object.

Distributed behavior is achieved by specifying the structural locations where we need to add behavior. This abstract statement is simply the goal of most AOP languages. Though it does not replace existing methods, only adding behavior. This makes it different to AOP languages like ASPECTJ where the `around` construct allows the bypassing of an existing method and actually redefine it. Similarly, COMPOSE* allows to superimpose a filter intercepting calls to an object and redefine the behavior associated with a given message therefore supporting bypassing of methods and redefining behavior associated with given messages. ASPECTJ is a general purpose AOP language, while DEMETERJ implements the concepts of traversals. Adaptive programming is considered a special case of aspect orientation.

Traits Related approaches such as *Traits* (see [69]) have not been explored in this work. Since Traits propose an alternative to multiple inheritance and do not contemplate a join point model. Schärli et al. design a mechanism to facilitate reuse of units considered in between classes and methods, in order to promote a fine grained reuse in object orientation. Schärli et al. motivate their approach as a solution to the conceptual problems of multiple inheritance. In short, they expose that classes play two roles: as *generators of instances* and as *units of reuse*. As instance generators classes should have a unique place in the class hierarchy, while as units of reuse they should be applicable at arbitrary places.

If we examine Traits as a solution to above conflict, we may prefer Traits to a more general mechanism as the "join point model." Traits represent a mapping of labels to methods and provide composition mechanisms to build classes using the mappings as building blocks. Comparing this approach to a solution with ASPECTJ, we may actually obtain a more fine grained class structure with Traits than altering the flow of a program as with the join point model.

After having introduced the techniques of aspect-oriented programming, in the following sections we propose some definitions that will clarify aspect orientation from a broader software engineering perspective in the coming chapter.

Chapter 3

Exploring the Concepts of Aspect Orientation

... shown by an examination of historical episodes and by an abstract analysis of the relation between idea and action. The only principle that does not inhibit progress is: anything goes. *-Paul Feyerabend, Against Method*

When a new concept appears, it seems necessary to answer at least the following two questions: what is it (definition) and what is it actually good for (benefit). When at the beginning of the doctoral studies the author was suggested[1] to do research on aspect orientation he did not have the slightest idea how difficult it would be to set terms straight in this field. Whether the message is addressed to an audience familiar to AO or not any answer to the first question gravitates around the programming level approaches to AO, particularly ASPECTJ, and the cluster of languages related to it. This renders the second question hard to answer and may even give reasons to argue against aspect orientation[2]. In most cases, prevailing concepts of AO are heavily influenced by a definition of aspect as a *join point and advice* entity; see Sect. 2.2. In some other cases, the answer to the first question is formulated with respect to another concept: *crosscutting concern*; see [51]. Both approaches may concur, consider for instance a programming entity aspect (i.e. jp + advice) representing the actual implementation of a *crosscutting concern*, see Example 2.2.4. Nevertheless, both do not necessarily coincide. Since we might write an aspect in ASPECTJ that acts upon one other entity without affecting third entities (we mean, modularization units such as objects). In such a case, we may clearly ask ourselves what the benefit of such an aspect would be. Here we come

[1]by Prof. Dr. Petr Kroha at the *Technische Universität Chemnitz.*

[2]The reader may refer to a master thesis in [64] that Tilman Seifert and the author supervised. This masters thesis outlines the limitations of the code level aspect constructs in a critical review of the subject.

to a third possibility, aspects as *optional* or *additional* functions on a so called base model (see also Example 2.2.4).

The clue we discovered is the following: the three possibilities mentioned above (see also Table 1.1) relate each to different abstraction levels (considered as phases or steps in the software development process). We would rather elaborate on what we consider the kernel of AO, namely *crosscutting* and *weaving*, both explored along this and the following chapters. We relate these three approaches to their corresponding development phase in Table 1.1.

This first classification helps in structuring our concepts in a more systematic way and advances an answer to the " what for" question.

We consider AO a technique to be useful in providing a modularization layer additional to the already existing ones. This is consistent with another useful characteristic of AO, namely supporting the insertion of additional or optional functional elements.[3]

It seems to us that the difficulties in accepting or rejecting some definitions are sometimes also " kind of" political. We leave aside such considerations. Our goal at this time has been to explore these concepts seriously and honestly.

More than that, as Karl Popper in his last book mentions, "ultimate knowledge is an empty word... science is the quest for truth.[4]" We hope to have gotten close to a fair and scientifically acceptable understanding of the topic.

As for the second question, one of the major and most disturbing difficulties to overcome is an obvious related debate: *is it not possible to achieve the same results with existing techniques?* Which may also be read as: do we *really* need aspect orientation? The answer depends, as mentioned before, on our definition of *aspect* and therefore the possible resulting techniques and methods we derive from it. Our choice has been outlined in the previous paragraphs.

We proceed as follows. First, we provide the context needed to support our definitions, namely, a system model and an example that helps illustrating the transformation of informal requirements into specifications and show an example of when these crosscut several modularization units. Second, we explore some possible classification criteria and introduce a taxonomy of aspects. Finally, we provide an answer to the second question: what is better with aspect orientation?

Exploring some preliminary classification criteria with respect to code, what we consider a bottom up perspective, aspects might be classified based on *pointcut* granularity as in the work of [30] but also according to the present main available

[3]see for instance Chap. 5. In that chapter we explain a composition framework for logical components. This framework can also be used to insert additional elements to a given model.

[4]"..denn Wissen -sicheres Wissen- ist ein leeres Wort. ... Wissenschaft ist Wahrheitssuche." [63]

AO technologies represented in programming languages; as shown in Table 2.1. Another criteria might also divide them in *Dynamic* and *Static* aspects, say from a systems analysis and design point of view or a programming perspective. For example, classifying by the moment at which *weaving*[5] takes place. Another possible classification might relate to the nature of the crosscutting expressed in the aspect language. We find an indication of the criteria we might rather follow in the work of Kiczales et al in [52]. The authors propose a definition of "Crosscutting" independent of a programming perspective, though it is actually an explanation of when a programming language is considered *aspect-oriented* relatable or not. However this definition hints at a tracing relation of aspects and requirements. This we take as point of departure to investigate the *nature* of the problem in crosscutting.

Regarding the above mentioned classification criteria, we consider that in the search for a thorough understanding of AO at large, they do not suffice. Mainly, because concentrating on the programming level restricts our understanding and would not allow us to conceive a scheme that explains (functional) AO as the one we will introduce in Chap. 4. For example, above all, considering that one of the main issues AO aims at solving is improving modularization. For modularization, we understand the process of obtaining independent pieces of software, or modules, that put together constitute a working system. We refer to modules independently of the step in the development cycle. For instance, we consider a clearly identifiable and independent model element as a module. Where a model is "an approximation, representation or idealization of selected aspects of the structure, behavior, operation or other characteristics of a real world process, concept or system" [37].

Criteria: Modularization Modularization relies on the concepts (in the sense of *mental frameworks*) available for software modeling, for instance, class, method, inheritance as in object-oriented programming languages. The design decisions are taken in the cumulative process of translating requirements up to code also may be considered as a collection of composition and decomposition steps. At the general level, composition of the concerns in a system with composition mechanisms that support the interaction of the entities pertaining to a system. At a more particular level, the decomposition of functionality as defined by Harel and Pnuelli in [31]. We consider that C^3 exist from the perspective of decomposition and modularization in a tracing relation with requirements and specifications. Both decomposition and modularization need a subsequent composition in order

[5]The concept of weaving at the programming language level is found in Sect. 2, we introduce it more formally in Sect. 4.2.3

to produce an operational software system. Therefore, *compositionality*[6] is a key concept in AO. This can be proven by the fact that the notion of *weaving* is present in every aspect-oriented technique, paper or related proposal.

In short The production of a brand new software, as well as the modification or maintenance of an existing one, relates to the requirements. Also in the case of non-preventive maintenance, since non-preventive maintenance deals with fixing of errors or failures stemming from changes in the environment or a wrongful implementation with respect to some desired functionality (implicitly or explicitly) represented by the requirements. In such case, the system is corrected in order to fulfill the desired functionality as stated by the stakeholders in the specification. C^3 are either stated in the requirements artifacts or are added by later design decisions. For instance, the requirements of the web store model we introduce later contain a precondition in use case 1 (see Appendix A.1.1) "The web store is ready to take orders. The article catalog is set, as well as prices and transportation costs." A straightforward reading of the precondition indicates writting some control function that validates three conditions:

- □ The article catalog is set.
- □ Prices are established.
- □ Transportation costs are determined.

Not only that, following a particular modularization structure as the one we obtain for the web store later in this chapter (see Fig. A.7), the function implementing this precondition would be defined over classes: Catalog, Article, and Office. All three above conditions are functional, for example, we may implement them as a boolean function checking that no article lacks price, and transportation costs are recorded.

As an example of C^3 from nonfunctional (or extra functional) requirements, consider a requisite establishing a given response time in use case 3 "Payment process" (Appendix A.1.3). This requirement may imply testing the system after constructing it to corroborate that from the time the client provides his payment information and the confirmation from the office to the client the time response is not exceeded. This condition is related at least to classes: Customer, Office, Bank, and eventually also Account.

[6] "...the meaning of a complex expression is fully determined by its structure and the meanings of its constituents - once we fix what the parts mean and how they are put together we have no more leeway regarding the meaning of the whole. This is the principle of compositionality, a fundamental presupposition of most contemporary work in semantics." [72]

Since aspects are a particular modular representation of C^3, we choose to classify aspects in view of its tracing from requirements instead of other possible classification schemes.

3.1 Current Concepts in Aspect Orientation

We now explore the subject from a code, i.e., bottom up perspective. We also provide a framework (see Sect. 3.2) at the hand of which we introduce our definitions; see Sect. 3.3. We consider that most common definitions of aspect fall into two groups. The first group consists of definitions stating that aspects are concerns that cut across other concerns (whereas two concerns crosscut if the methods related to those concerns intersect. This is basically the Kiczales et. al definition [19]). The second group assuming a very pragmatic approach and defining aspects in terms of the constructs of aspect-oriented programming languages: an aspect is "a join point and advice." A *join point* is defined as "a well defined point in the execution flow of the program", and an *advice* as "the actual implementation of the aspect." Weaving is the mechanism that actually inserts the advice at the points in the execution flow indicated by the join points. The first group argues that aspects produce tangled representations that are difficult to understand and maintain [65]. Moreover, crosscutting is relative to a particular decomposition [19]. We subscribe the definition to a certain extent yet we distinguish aspects from C^3. See for instance Fig. 3.1 which corresponds to a graphical representation of an aspect as compared to modular units.

In any case, aspects can be system properties involving more than one functional component thus crossing the static and dynamic structure of a program. Aspects affect and modify several classes. This means that aspects might be static or dynamic properties that affect not only the behavioral definition of the main modularization entity (e.g. class, component, service) but also its data structure. Most work in this field supports the idea that AO is an enhancement of object-oriented programming. AO is an enhancement relative to the main modularization paradigm.

Some authors define as C^3 issues that are not well localized in functional designs, such as: synchronization, component interaction, persistence, security control, fault tolerance mechanisms, quality of service and others; therefore these are concerns that are typical candidate aspects. Another guideline that some authors propose to identify aspects is that aspects "are usually expressed by small code fragments scattered throughout several functional components"; see [17]. Indeed compatibility, availability and security requirements are crosscutting concerns according to [66]. Similarly, exception handling, multi object protocols, synchronization and resource sharing would be extended across the source code if only using traditional implementation techniques [49]. However, we should discern between systemic and localized C^3 (or semi localized) which we will discuss in our aspect classification.

Fact: Although the above ideas provide material for a first definition of C^3, they still leave room for interpretation and do not help discerning C^3 from aspect.

Problem As Ossher in [19] mentions, "One of the hardest things about crosscutting concerns is understanding just what cuts across what". There is a need for research on aspect identification. Our work contributes to make AO more precise.As Kiczales, in [53], points out a simplified rule of thumb for aspect identification: " A property that must be implemented is

- □ a component, if it can be cleanly encapsulated in a generalized procedure (that is object, method, procedure, API). By cleanly, we mean well localized, and easily accessed and composed as necessary. Components tend to be units of the system's functional decomposition, such as image filters, bank accounts and GUI widgets,
- □ an aspect, if it can not be cleanly encapsulated. Aspects tend not to be units of the system's functional decomposition but rather to be properties that affect the performance, or semantics, of the components in systemic ways. Examples of aspects include memory access patterns and synchronization of concurrent objects."

Our critique, this simple "rule of thumb" points us in the right direction because it is consistent with the notion of C^3. Our critique to this simple rule is that it does not help discriminating what is and what is not an aspect. For instance, would a synchronization protocol be an aspect or would only a part of the protocol be considered an aspect? Or, would every call to a database in a front end be considered an aspect just because it might be carried out in different parts of the front end?

In other words, we consider that aspects acquire meaning at the level of the semantic dependencies in a software system whether they are constraints affecting more than one functional unit or user concerns that are implemented by more than one functional unit.

In search for meaning, it is commonly accepted that *aspects* are "concerns that cut across other concerns". According to the early notions of aspects, two concerns crosscut if the methods related to them intersect [20]. Therefore, C^3 are the same as aspects and produce tangled representations that are difficult to

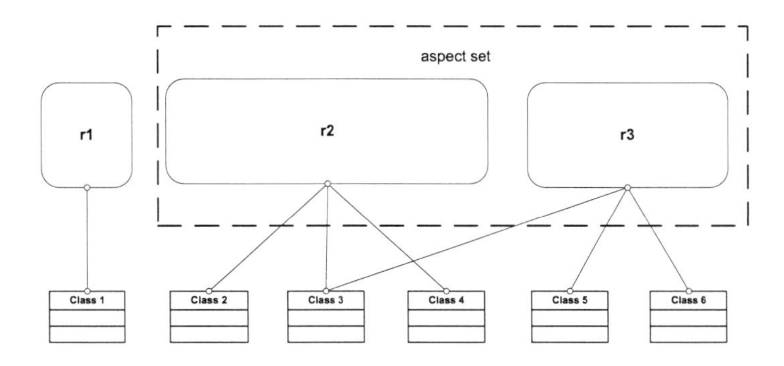

Figure 3.1: Requirements and aspects in relation to classes (in [24])

understand and maintain; see [67]. This crosscutting is relative to a particular decomposition; see [19]. A concern is considered as any area of interest in a software system.

These concepts are too general. That is why we propose our concepts in Sect. 3.3. In order to achieve it, we explore the meaning of the word aspect and reflect over the accepted notions of AO.

A little bit more philosophical. The word aspect has the following meaning [1]:

> *a particular status or phase in which something appears or may be regarded*

Besides, the word aspect comes from the latin *aspectus* or *aspicere*, that means " to look at." The selected definition and the etymological meaning of the word reflect the idea of a perspective towards an artifact and in this case towards a system. The word aspect is then consistent with conceiving a system as composed of different parts (subsystems) or angles, and allows us to separate such subsystems as areas of interest.

Fact nevertheless indicate that we must place emphasis on the fact that aspects are not simply "views" of a given system, unless some weaving mechanism supports merging the views.

Moreover, aspect identification relates to the need of an integral comprehension of concern crosscutting altogether with its context. As Videira Lopes ([79]) explicitly states: the problem AOSD solves is one of complexity in today's software applications.

3.2 A Framework for our Definitions

We use the concepts of *Problem Space* and *Solution Space*, or in the language of [41], the phenomena of the Problem World (PW, also called Problem Space or PS) and the phenomena at the interface PW- Machine (also called Solution Space), to examine the way in which aspects originate. We provide a formal example that explains the way in which C^3 originates from the design decisions and modularization concepts available to transform requirements into models and finally code. On the one hand, we introduce the requirements of the example case, the web garment store in Sect. 3.2.1; on the other, we elaborate the specifications and explain how we obtained them. The former represents the problem space and the latter the solution space.

We consider that the selected framework, namely problem frames, allows discerning requirements (in the problem space) and specifications (in the solution space). Since we relate aspects to a tracing relation in terms of requirements and later development phases, this distinction is valuable to clarify our definition and classification of aspects.

We express requirements in the form of use cases and their corresponding specifications as signatures, facts and predicates in the language Alloy [39].

3.2.1 An Object Model: Web Store Design

The on-line garment store we present here was first developed by BOC Consulting GmbH. A modified version of this web store by [38] was used for an application of a security analysis. We later referred to Ioshpe's version in [27] and introduced it as a case study for a formal model of composition of security aspects. The web store model, in [38], is already at the design phase of development. In this Work, we complete and modify the model by adding the system requirements in the form of use cases. We also elaborate a new structural diagram.

On the whole, we complete the case study with a system model consisting of use cases, a structural diagram (class diagram) and a formal model representing the structure and behavior of selected parts of the system.

The system model is inspired by the following diagrams as well as on an actual web store: www.dress-for-less.de:

- □ Class diagram (former version, we propose a different structural diagram in this work),
- □ deployment diagram,
- □ activity diagrams: *Bestellung* (Order), *Beschaffung* (Purchase), *Lieferung*

über Pick Point (Delivery to pick-point), and *Direkte Lieferung* (Direct delivery),

- sequence diagrams: *Beschaffung* (Purchase in two scenarios), *Bestellung* (Order in four scenarios), *Lieferung* (Delivery in three scenarios)

We support our use cases on the existing diagrams and obtain the formal model out of the use cases.

System Description Our web store is intended to reflect the behavior of most online businesses. It allows clients to buy articles (garment) on-line. The client may navigate a catalogue, select goods and purchase them. The client indicates the shipping address, pays by credit card, the money is transferred to the online store and the goods are sent through a shipper. The store manager sets the catalogue selecting garments from the offerings that the wholesaler presents in its corresponding inventory (also called warehouse catalogueing). The prices are settled by the base price from the wholesaler plus a given margin determined globally by the store manager together with the transportational costs. An overview of use cases and actors is shown in Fig. 3.2. The use cases are included in Appendix A.1.

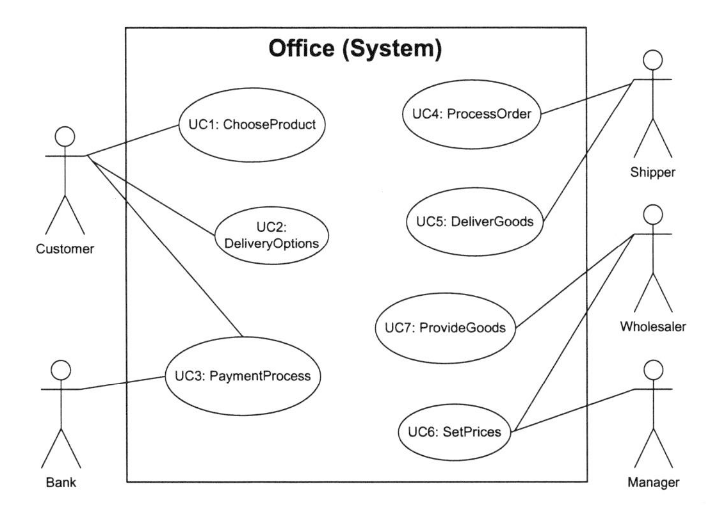

Figure 3.2: Web store. Overview of use cases

Obtaining the Structural Model (Class Diagram) from the Use Cases we deduce the system's structure and functionality from the use cases. Our method is straightforward. We take the actors and entities as candidate classes. Based the actions, as mentioned in the standard process of each use case, we draw the corresponding relations and confirm which candidate classes appear as classes in the diagram. These are basically design decisions. It is actually not the aim of this work to propose a software development methodology, we rather rest on the methodology of object orientation (as in [81]), and common practice.

We consider each a partial diagram as part of the whole and iterate once a subsequent use case refers to an already considered actor or entity. In each necessary case, the diagram is modified accordingly.

Use Case 1 We associate actors and entities (nouns) to preliminary classes. Afterwards, we revise the resulting diagram against the relations described in the use case among classes. Thereafter, we look for class attributes in the form of nouns and qualifiers. The process is reiterated when a common instance in a subsequent use case contradicts, or enhances, the existing partial diagram.

For instance, the actor customer is mentioned a number of times in the use cases. In Sect. A.1.1, we observe that the customer buys (relation) from the office (web store) by selecting articles and placing them in a shopping cart. Given its kind of autonomous existence, the shopping cart is made into a class. Also in this use case, garment (article) selection is completed by choosing: size and color. These we incorporate as properties of article so we make them to attributes of this class. The resulting diagram is shown in Fig. 3.3.

We progress similarly along the seven use cases and reiterate whenever a class, relation, or attribute, is contradicted or enhanced. This process terminates when the last use case is translated into a partial class diagram and then we elaborate upon the complete structural diagram by assembling the partial diagrams together.

The behavior is obtained in another round of iterations. These are written in the form of specifications together with the structure, particularly as signatures, predicates and facts in Alloy.

The other use cases, and their transformations, can be found in Appendix A.2 where the structural diagrams and transformation steps are found. We completed the structural model by setting the partial diagrams together in Fig. A.7, in each necessary case, we would iterate again.

Specifications of our web store model is built upon a relatively easy to understand formal language supported by predicate, relational and "navigational" calculus. The objective is to provide a case study on which we may illustrate a premise. Namely that crosscutting concerns are requirements formulated in the problem space. Also, that after being translated into concepts of the solution space (e.g., modularization units such as objects or logical components), can be traced to more

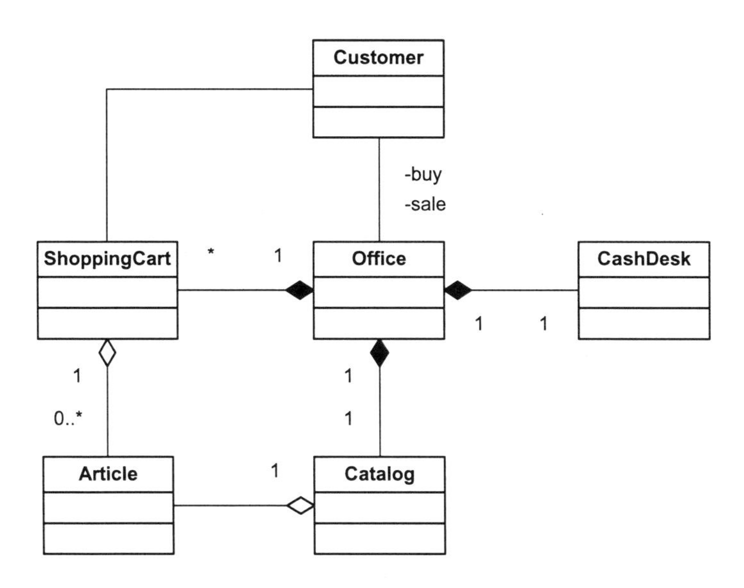

Figure 3.3: Classes from UC1

than one such unit in the solution space. Following this case example, aspects are defined (see Definition 6) as an additional software abstraction or concept that supports modularizing such crosscutting concerns.

The Formal Language: Alloy

We recall this subsection from [40]. All structures in the lightweight formal language Alloy are built from atoms and relations. Atoms are the basic entities and the relationships are built among them. An atom is a primitive entity; it is *indivisible*, *immutable* and *uninterpreted.* In order to model something that is divisible, mutable or interpreted, we need to specify relationships. A relationship is a structure that relates atoms.

At the core of this modeling language is a small and expressive logic. It is a relational logic that combines the quantifiers of first order logic with the operators of the relational calculus. A particular characteristic of Alloy is its generalization of the notion of relational join. In short, a relationship is a set of tuples; sets are represented as relationships with a single column and scalars as singleton sets. Henceforth, the same join operator can be applied to scalars, sets and relations.

Alloy's logic supports the following styles: predicate calculus, navigation expression and relational calculus.

Class Diagrams and Alloy Models Now, we draw the relation between our structural diagrams (class diagrams) and the Alloy model we build upon it. From the highest to the lowest, we have the following levels of abstraction in this language: the Object-Oriented paradigm, set theory and atoms and relationships. The last one resembles the true semantics of the language.

We translate the classes obtained from the use cases, Fig. 3.3 through Fig. A.6, into a model whose syntax and semantics are provided for by the language Alloy.

The structural specifications can be found in Appendix A.3.

3.3 Transformation Example and Definitions

The exercise, or case study, realized here, transforms the web store design use cases into specifications, provides an example to explore the source of crosscutting in the translation from requirements to specifications (or design). Namely, in the transformation from a *problem space* to a *solution space* which is where we may identify requirements that crosscut modeling units such as classes or specifications. Through this exercise, we have achieved the following schema (illustrated in Fig. 3.4) that illustrates two behavioral specifications: one crosscutting several classes and the other affecting only one.

We will introduce now the concepts which provide the underpinnings of our classification. As already mentioned, in our model of the world (actually of software development), software requirements relates to a *problem space*, in the early requirements stage and are mapped to a *solution space* in the later requirements and design stages.

> ...the expression "requirements specification" by itself is virtually meaningless. Whenever we use this term, it refers to a deliverable of development consisting of product objective and of required product behavior [82].

Definition 1 ***(Requirements)*** *We recall the definition of requirements of Ian Sommerville [71]: "User requirements are statements, in a natural language plus diagrams, of what services the system is expected to provide and the constraints under which it must operate."*

Definition 2 ***(Problem space)*** *The problem space is defined as the set of requirements together with the explicit, or implicit, definition of the environment in which how the system to be should operate.*

We provide an example of the problem space as use cases in Appendix A.1, illustrated at the top of Fig. 3.4. Another example of problem space, related to an e-learning support system, is provided in Appendix D.

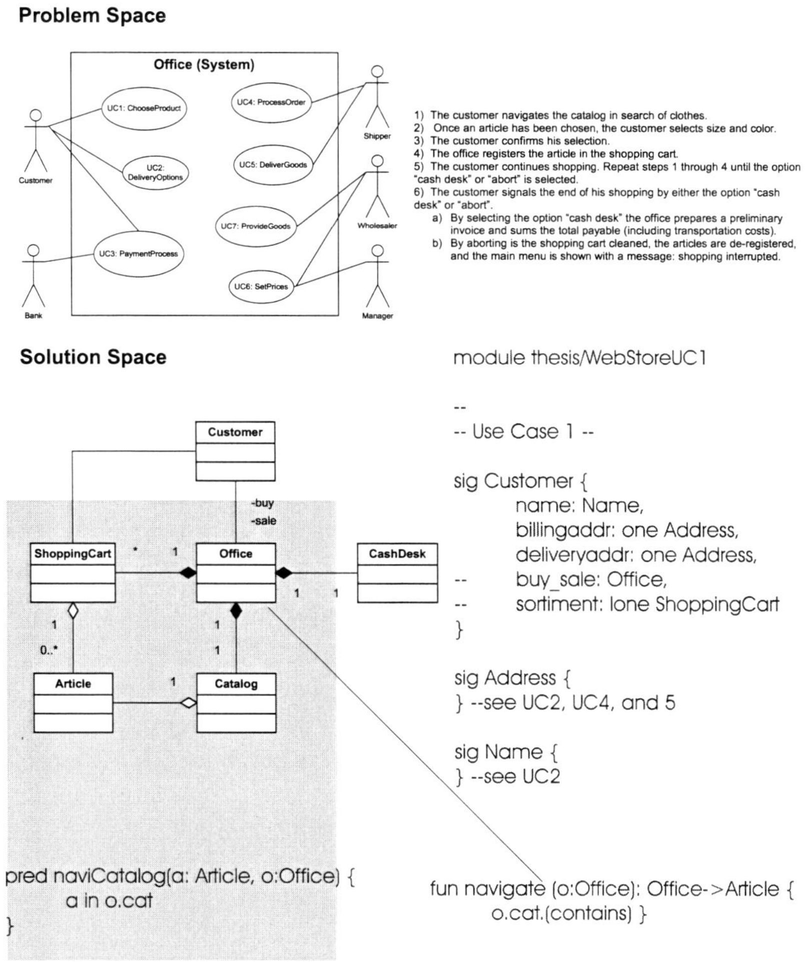

Problem Space

Office (System)

UC1: ChooseProduct
UC2: DeliveryOptions
UC3: PaymentProcess
UC4: ProcessOrder
UC5: DeliverGoods
UC7: ProvideGoods
UC6: SetPrices
Customer
Bank
Shipper
Wholesaler
Manager

1) The customer navigates the catalog in search of clothes.
2) Once an article has been chosen, the customer selects size and color.
3) The customer confirms his selection.
4) The office registers the article in the shopping cart.
5) The customer continues shopping. Repeat steps 1 through 4 until the option "cash desk" or "abort" is selected.
6) The customer signals the end of his shopping by either the option "cash desk" or "abort".
   a) By selecting the option "cash desk" the office prepares a preliminary invoice and sums the total payable (including transportation costs).
   b) By aborting is the shopping cart cleaned, the articles are de-registered, and the main menu is shown with a message: shopping interrupted.

Solution Space

Customer
-buy
-sale
ShoppingCart
Office
CashDesk
Article
Catalog

```
module thesis/WebStoreUC1

--
-- Use Case 1 --

sig Customer {
        name: Name,
        billingaddr: one Address,
        deliveryaddr: one Address,
--      buy_sale: Office,
--      sortiment: lone ShoppingCart
}

sig Address {
} --see UC2, UC4, and 5

sig Name {
} --see UC2
```

```
pred naviCatalog(a: Article, o:Office) {
        a in o.cat
}
```

```
fun navigate (o:Office): Office->Article {
        o.cat.(contains) }
```

Figure 3.4: Source of crosscutting: tracing from requirements to specifications

Definition 3 *(Solution space) The solution space is the set of artifacts that belongs to a running implementation of a software system and without which the system would not be able to operate or would not have been produced (requirements are explicitly not in this set).*

In the web store example, the solution space consists of structural diagrams (see Fig. 3.3 through Fig. A.6, Fig. A.7 as well as the specifications in Appendix A.3). Another example of solution space consists of the artifacts from the ELSS; see Appendix D.2.

Definition 4 ***(Transformation from problem space to solution space)*** *Given the notion of refinement as the transformation of an abstract, i.e., high level specification into one or more concrete or low level executable software artifacts, we defined the transformation from problem space to solution space as the process of transforming a given requirement into one or more executable artifacts by means of refinement. This mapping is performed explicitly, or implicitly, with the help of a given conceptual model that translates requirements to software models and finally to code.*

Moreover, solution and problem spaces are illustrated in Fig. 3.5 in a more general way than in Fig. 3.4. The transformation is represented by the arrow from one to the other. We may formally relate the mapping either to a transformation function from problem to solution space; Sect. 3.2.1 provides an example of such transformation.

Requirements give shape to an architectural style, platform, interfaces, etc. in a certain domain and following the conceptual model in use. In short, the translation process from the problem space to the solution space is performed using some given conceptual models, e.g., object orientation, components, service-oriented approaches, and other relatable areas.

Some requirements are implemented in several modularization units. This suggests a way to define crosscutting as a functionality or constraint characterizing more than one (logical at design, physical at code) module from one abstraction level to the next. This can be better understood by considering an abstraction level as a refinement step in a series of steps from requirements to code, being the previous step the most abstract in relation to the next one, such a relation is clearly transitive. As a matter of fact, we may build a more abstract definition of C^3 than the ones explored so far in this section by considering behavior (functionality) as a property[7] implemented in several modules. In other words, *more than one* modularization entity is present. The modularization entity depends on the chosen architectural paradigm, e.g., object orientation, components, agents, etc.

The above reasoning leads to the following definition of crosscutting concern.

Definition 5 ***(Crosscutting concern)*** *Given a problem space a C^3 is a requirement that under some translation from the problem space to the solution space is expressed in more than one modularization unit in a lower level of abstraction.*

We differentiate C^3 from *aspect* as follows. Crosscutting is due to the translation from the problem space to the solution space (Fig. 3.5) given the fact that

[7] A property is formally defined as a set of behaviors, so that an execution of a system Π satisfies a property *P* if and only if the behavior (a sequence of states and agents) that represents the execution is an element of *P*. [2]

no modularization abstraction is perfect while aspects have accomplished an additional modularization unit to implement such crosscutting in models or code. Emphasizing this idea, aspects exist at the software architecture, design and implementation stages whereas C^3 can be seen as a superordinate concept, i.e., more generic concept. This makes the aspects a subset of crosscutting concerns.

Definition 6 ***(Aspect)*** *Given a solution space an aspect is a specification (in the formal language Alloy: predicate, function or fact) that represents a module with behavior crosscutting other modularization units (e.g., class, component, function); with respect to the underlying architectural framework.*

In our web store example, aspects are specifications (Alloy functions, facts, and predicates) that are related to (in sense of *operating over*) a set of classes. As illustrated in Fig. 3.4, the predicate `naviCatalog` (see Listing 3.1) affecting classes `ShoppingCart`, `Office`, `Catalog`, and `Article`. In contrast to function `navigate` in Listing 3.2.

Nevertheless, we would like to emphasize the idea that the entity representing, or implementing, the crosscutting does not necessarily have to be a specification. It may be represented as an additional construct over the base modularization units. For instance, the composition filters defined as logical components and inserted over the communication channels between components illustrated later in Fig. 5.1; see Sect. 5.2.

We introduced the current concepts of AO, identified its main concepts, introduced our definitions. In the following, we identify some types of C^3, characterize the sources of crosscutting, and present our classification of aspects.

Functional vs. nonfunctional C^3 Requirements may be classified in *Functional (FR), Nonfunctional (NFR), Design and Implementation.* We understand FR as defining characteristics of the problem space[8] and NFR as constraints in the solution space[9]. The behavior of the SuD is expressed as FRs whereas restrictions to the possible solutions are defined in the form of NFRs or Design Requirements. Nevertheless, nonfunctional requirements are at some point in the development cycle translated into functional specifications. Meaning that they are translated into quantifications or behavior. As an example, consider a NFR such as "Fault Tolerance". This requirement can, in subsequent phases of development, be translated into functional specifications that guarantee data persistence in view of a system failure. We group C^3 in Functional and Nonfunctional. This will help us constructing our classification of aspects. In Definition 6 we introduce a definition

[8]What a system should be able to do, the functions it should be able to perform.

[9]Requirements on the system's performance. For instance, the amount of users that have to be attended in a given time frame.

Listing 3.1: Predicate navigate catalog

```
pred naviCatalog(a: Article, o:Office) {
        a in o.cat
}
```

Listing 3.2: Function browse catalog

```
fun navigate (o:Office): Office->Article {
        o.cat.(contains)
}
```

of aspect that can also be formulated as "a module implementing a functionality that modifies the functionality of other modules and is back traced to a C^3." We therefore use the distinction FR and NFR to succinctly categorize aspects.

3.4 Sources of Crosscutting

3.4.1 Crosscutting Due to Inherent Limitations in a Decomposition Paradigm

One source of crosscutting is due to the limitations of existing programming as well as design paradigms. As an example, consider object orientation. The building block in object orientation is the class, which consists of data and methods. Some functionality mentioned in the requirements may be common to several classes. The methods implementing such requirement are spread out amongst several classes. For instance, the predicate specifying the navigation of the catalog in our web store example. This predicate affects classes: *Office*, *Catalog*, *Article* and depending on other design decision, also *ShoppingCart*; see Fig. 3.4. Compare this to a function for the customer to provide his name and address This function operates only on class *customer*.

In other words, systemic concerns such as those that relate to a group of classes (e.g. some security concerns like access control) are implemented into methods of several classes. For example, see Table 3.1, in this table, we show some preconditions and post conditions that are in relationship to several signatures, as opposed to the use cases shown in Table 3.2 which are in relationship basically to only one signature.

We may think of this as a set W containing all the methods that realize access control. To further illustrate this, assume we have classes `client`, `server`, and `audit`, more on we write methods `client.authentication`, `server.-`

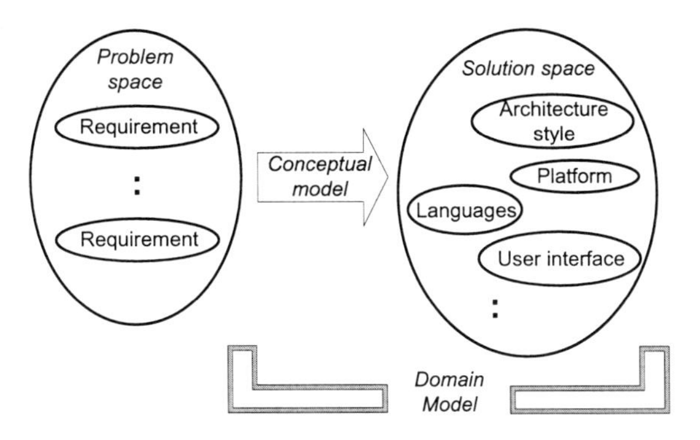

Figure 3.5: Problem space and solution space

`authentication,...,server.authorize.`

$$W = \{client.authentication, server.authentication, \ldots, server.authorize\}$$

The reason for systemic concerns is that no programming paradigm is perfect in the sense of allowing for a "concern-oriented" modularization. Given the fact that the abstractions shaping a software system are constrained by the underlying modeling or implementation paradigm. For instance, the class/object concept, or any other.

Indeed, modularization abstractions cause C^3. In the case of object orientation this is explored in [4] and more so in depth by S. Clarke in [12]. Particularly, S. Clarke demonstrates that the units of modularization in the OOP are structurally different from the units of modularization in requirements specification. This result can be generalized to other modularization paradigms. We express this notion in Definition 5.

3.4.2 Crosscutting from the Transformation of Non Functional Requirements into a Functional Proxy

Another source of crosscutting lies in the translation of NFR into a functional implementation or a functional proxy of the corresponding NF specification. These are also considered in Definition 6 through the *Implements* relationship. For instance, in [58] we present the translation of the security constraint *Keep transaction secure* in the early requirements stage into a number of security (sub)constraints such as *keep transaction private, keep transaction available* and *keep integrity of the transaction* in the late requirements stage. These (sub)constraints represent the functional proxy of the more abstract security requirement.

Table 3.1: Some examples of functional C^3 from the use cases of the web store

	Concern i.e.informal specification	Signatures
precondition UC1	(article catalog set) (prices set) (transportation cost set)	Catalog Article Office
post condition UC1	(shopping cart not empty) (preliminary invoice ready)	ShoppingCart CashDesk Office
post condition UC2	(shopping cart not empty) (Customer.name, .billingaddr, .deliveryaddr are given)	ShoppingCart Customer

Table 3.2: Some examples of non C^3 from the use cases of the web store

Use Case	Specification	Signature
UC1. Standard process	customer selects articles	ShoppingCart
UC1. Standard process	customer confirms selection	ShoppingCart
precondition UC2	(shopping cart not empty)	ShoppingCart

As already mentioned, crosscutting concerns appear in the transformation from the problem space, where requirements are formulated, to the solution space, where the architecture, deployment, etc. are defined. Take for instance, fault tolerance. Guaranteeing data persistence despite unanticipated faults may require a set of functions that are implemented in several modularization units (components, classes, services).

Design requirements relate to decisions regarding architectural style, user interface style and decisions that constrain the set of options available in the domain model. The domain is also referred as *Universe of Discourse* (UoD) in the literature. It is used to mean the part of the world to which the data manipulated by the *System under Development* (SuD) [10] refers. We consider adaptability and

[10]"Because computer-based systems manipulate data, there are two systems we can specify: the

maintainability as design requirements.

Performance requirements for instance, usually considered nonfunctional, are namely translated in specifications that precisely quantify what performance actually means for the *SuD*. Same requirements occur in the case of security requirements. Components performing security protocols such as encryption, authentication, etc. can be considered proxies for more abstract security requirements, e.g., data integrity. These proxies not only act upon several modularization units in the resulting system, but are also implemented in several modularization units.

What we get from this As a result of this analysis and definitions we come to the following conclusions:

1. Crosscutting can be described in terms of a tracing relationship from requirements to system design as well as from this to code. It is a transitive relationship.

2. An aspect is the representation of a crosscutting concern. One possible way to describe aspects can be a special refinement function, namely communication refinement. We recall this definition in Chap. 4.

3. Crosscutting is the outcome of the translation from a higher to a lower abstraction level starting with requirements and ending with code.

4. Crosscutting is due to the limitations of existing modularization concepts as well as design decisions; and it will always exist.

5. Crosscutting is determined by the modularization technique and is given after requirements translation to further steps in the development process.

6. C^3 may be managed in design and code by means of *aspects*. We explain this in Chap. 2 and 5.

3.5 A Classification of Aspects

We classify aspects based on a classification of requirements since we define them as implementations of crosscutting concerns. Therefore, we draw a parallel between requirements and aspects with respect to this classification which we outline in Table 3.3.

We draw our examples from the web store case previously introduced and further contained in Appendix A. We also draw some examples from a prototype for

computer-based system and its UoD. In a development process, the computer-based system is also called the **system under development** or SuD" [82].

an E-Learning Support System (ELSS) we wrote and whose technical report can be found in [23]. The ELSS was produced by the author in the context of a Data Mining and Warehousing course. We recall parts of the system in Appendix D.

Example 3.5. The E-Learning Support System The system is designed to facilitate the evaluation of students, which are supposed to attend a number of courses. Some of these courses contain a number of self study hours. At the end of a section students must apply an examination. The lecturers set the contents of the tests, the system then provides a front end for students to answer. The front end also allows lecturers setting questions together with the correct answers, and the system supports evaluating students based on the given answers. Therefore, students get feedback right at the end of the evaluation. The director of the center may check on the statistics to have an overview of the students' performance. The ELSS was designed as an n layer system, this means it is structured into: front end, application layer and database layer. A broader description of the ELSS, specially the requirements can be found in Appendix D.1.4.

3.5.1 Nonfunctional Aspects

We mentioned before that a NFR like *Fault tolerance* is subsequently translated to more concrete specifications that may finally affect a number of entities, think about a component or set of components guaranteeing data persistence in view of a system failure. More specifically, at the case of the ELSS we might have a set of routines to ensure commiting information (student's answers in a test, the questions entered by the lecturer, etc.) to the database once entered. Considered broadly, the implementation of such a requirement is dispersed through the three layers of the system yet predominantly in the database manager and the in application layer. In case we may draw a clear borderline of the entities affected, we identify nonfunctional aspects as systemic (NFS) or semi localized (NFSL).

NFS are aspects that given a boolean function "operates on" of the form:

Table 3.3: A classification of aspects

Nonfunctional	Systemic	NFS
	Semi localized	NFSL
Functional	Systemic	FS
	Semi localized	FSL

$$OperateOn : Component \times Aspect \mapsto \{1, 0\}$$

Where component stands for modular entity. The function gives true (1) in case the component implements the aspect at least partly, and false (0) otherwise.

NFS We define a NFS as an aspect that stems from a NFR and the function OperateOn gives true for more than half of the modules. In the case of the above example, we equate modules to layers. For example, "fault tolerance" implemented by a set of routines related to all three layers of the system and therefore represents a systemic nonfunctional aspect.

NFSL We define a NFSL as an aspect that stems from a NFR and the function OperateOn is true for less than half of the modules. For instance, users of the ELSS may access the system remotely, a requirement like keep information secret to third parties demands adding an encryption protocol between front end and application layer yet we may consider it is set only at the communication links, not affecting other parts of the system.

3.5.2 Functional Aspects

Aspects that stem from functional requirements, or that originate from the operationalization of nonfunctional requirements, can be formulated in terms of behavior rather than constraints and constitute Functional Aspects (FA).

We explain these by the following example. The subrequirement "keep transaction secure" is derived from the requirement as "Student information should be kept confidential, specially grades"; see Appendix D.1.5. This requirement can be implemented by an encryption protocol as the one we explore in Chap. 5. This aspect is in relationship to several other modularization units. We may also define a function OperateOn to distinguish systemic from semi localized functional aspects, following the extent to which these affect the base model.

Moreover, FA's can be implemented at the level of the methods in the classes, or at the level of behavioral specifications in components. For instance, black box aspects are related into the public interfaces of components like functions, object methods, and communication channels. The behavior affecting several modularization units can be related to these via some wrapping over the communication channels or the external interfaces. Such as those that are capable of being implemented by frameworks as Composition Filters. We explore these aspects in Chap. 5. Moreover, clear box aspects relate to the internal structure of the classes or components. In this case, the aspects can be seen as additional units composed inside the base ones. Ivar Jacobson introduces these aspects, years before the concept of aspect orientation, as "Existion and Extensions" in [42] illustrated in Fig. 2.1(b).

We propose a composition model for FA in Chap. 5.

3.6 Related Work

A review of other classifications reflects the predominant role of the programming level approach towards aspects. See for instance the work of [13]. The author proposes that aspects be categorized into two sorts: “spectators” and “assistants” with relation to the behavior of the code that they advise. This classification is certainly of interest for programmers. Moreover, [68] classify aspects based on the interaction between advice and method. Their classification helps programmers understanding the possible interactions of a given aspect and its design implications.

There is a categorization of aspects at the programming level related to classes of temporal properties in [48]. The author defines the following classes of aspect: spectative, regulative and invasive. We share a common interest in providing aspects with semantics and our work in Chap. 5 also explores aspect interaction yet it is at a different abstraction level. It is nevertheless worth noting that we also chose state machines to provide aspects with semantics in the above mentioned chapter. But at the same time, our classification in this chapter and the modeling framework for functional aspects in the coming chapter do not describe aspects in terms of temporal logic properties rather in terms of aspect weaving and its effect on the communication service (further explored in Chap. 4).

We aim at a more comprehensive and top down aspect classification. In this way, though still very much at the language level, the work of [30] is quite comprehensive. The authors classify aspect-oriented systems based on type of join point, selection criteria and the adaptation mechanism. They classify the main aspect-oriented languages based on the above criteria. Their work helps selecting the aspect language, or weaving mechanism, more suitable for the implementation of a given crosscutting concern. [56] propose a framework based on crosscutting concern sorts intended for aspect mining techniques. These sorts are defined as atomic descriptions of crosscutting functionality. Classified based on “intent” and its relationship to an aspect mechanism (at the programming language level). We agree that the authors present a classification that supports aspect mining though very much influenced by the actual programming level constructs. This means that the elements they mine might not necessarily cover aspects in general though certainly those that they give *a priori* to are defined as C^3. In contrast, we define crosscutting and aspect and provide a classification that is independent of language mechanisms.

We presented early work on a taxonomy of aspects in December 2005 at the University of Twente. As a result of the discussion, it occurred that the author's proposal was justified in the terms of "requirements traceability." At that time, we had already published a definition of aspect and its relation to a formal theory for modeling aspects (see [24]). In that work, we related aspects and requirements to traceability. Similar work from [76] discussed crosscutting, though in terms of *tangling* and *scattering*, which are not yet not related to requirements traceability. Later on [78] relate their definitions to requirements traceability, although they propose no classification and focus on identification of crosscutting in the early phases of software development. Our first exploration of the ideas introduced in this chapter can be found in [25]. In this chapter, we improved the definitions and enhanced previous work with concrete examples. Particularly in the area of the web store system model that introduces a transformation example from requirements to specifications that was needed to illustrate our concepts. We also obtained a more clear classification of aspects by categorizing based on the type of requirement from which an aspect stems and the length to which the aspect relates to the system.

3.7 Summary

In this chapter, we provided a systematic study of aspect orientation from a *top down* perspective. We explored the genesis of C^3 and aspects. Moreover, as a consequence of relating their causation to the transformation from a problem to a solution space, we grouped them based on their *nature*. In other words, we may then classify them based on the classification of requirements in functional or nonfunctional. We enrich this first level of classification with a criteria on the extent to which the aspect affects the base modularization units, so we obtain a second level of classification in: systemic or semi localized.

The definitions introduced here and the classification of crosscutting concerns in functional and nonfunctional may help identifying the technology at hand for dealing with each type of aspect. This is achieved by proposing a modeling approach for functional aspects in the coming chapter. We are in complete agreement that further studies are needed to relate this type of aspect to the corresponding design and implementation technology in a more integral way. Although it is not one of the goals in this research.

The results of this work might assist identifying the software processes at which the application of AO actually represents an improvement and discard those in which it might make less sense. This work seeks to provide more clarity for the coming phases of *aspect orientation* as it moves from being a maverick code related idea to a more consolidated and integral field of research.

Chapter 4

A Formal Approach to Aspect Weaving

You need to have a good technical foundation in order to be able to throw everything overboard and get through to the essence.
-André Sollie, painter

In this chapter, we propose that weaving of functional aspects can be explained by the notion of communication refinement. We defined the concept of communication refinement in Sect. 4.2.2 and equated it to *weaving* of aspects at the modeling level as opposed to aspect weaving into code. In this regard, models are viewed as sets of components interacting via some means of communication, e.g., connection oriented or connectionless communication.

Communication refinement is here portrayed as a kind of behavioral refinement that helps adding functionality over a given set of components at the communication channels between the components, i.e., adding behavior to the base model. Communication refinement is applied to the communication service. This way of adding behavior opens up a way to stepwise add layers or sets of new components to the previously existing ones. This is how we explained *aspect weaving*. Still, communication refinement itself does not explain the selection mechanism or join point model, we introduce a composition model in the next chapter in which we indicate through some tables the composition order and components amongst which the aspects are introduced. The security exampls in the coming chapter provide a more applied approach towards the concepts we introduced here.

In Chap. 3, we explored the actual concepts used in AO, defined our concept of crosscutting concern, aspect and introduced a classification. Our classification considers two main types of aspects classified in relationship to software requirements: functional and nonfunctional aspects. In Sect. 4.1, we provided an introduction to the mathematical formalism we use to model *functional aspects*.

In the following, we will provide the underpinnings to analyze *black box* functional aspects. *Black box* emphasizes the fact that we quantify our reasoning over the public interfaces of components. In contrast to that, *clear box* aspect orientation quantifies over the internal structure of components; see also [21].

In general terms, we may choose several focal points with this formalism. We may choose components as the central analysis entity, communication channels or interfaces. Whereas interfaces are made up of (typed) communication channels, and components have input and output interfaces. By selecting communication refinement to explain weaving, we portray refinement at the level of communication channels. However, the formalism is powerful enough to allow us for also considering subcomposing or refining components themselves at which we may further weave additional components. To explain this, consider Fig. 4.1. The interface consists of input channels $s_1, \ldots, s_i, \ldots, s_n$ and output channels $o_1, \ldots, o_m$. Weaving at the channel level means that we select one or several of the channels and turn these into a "first order" entity. In the first place, these channels do not process messages themselves, any process is performed in the component level. However, when we refine any of the communication channels it is as if we add one or more components between the original sender and receiver to which the channel belongs. This is one of the focal points we mean above. The other focal point is centering our attention to the component, e.g., $\mathcal{P}$ in Fig. 4.1, and introduce one or more additional components. This is further explored in Chap. 5.

We recall the formal definitions.

4.1 Formal Foundations

In this section, we recall the mathematical foundation to formally describe aspect orientation. We call the smallest functional entity, of which a model is composed of, a *component*. This approach is neutral to, for example, object orientation. The reason is that AO is a composition technique and in this regard not bound to object orientation or any other modern development paradigm.

In this formalism, components communicate over channels via streams of messages. Streams, manifesting a complete communication history, make it very convenient to define and reason about relationships of composition. The definitions we introduce in the following as well as the formal and notational conventions are those of FOCUS [11]. This chapter is inspired on early work that related the subject to a formal theory and was published in [24]. This work was later enriched through discussions the author had with Prof. Broy and Prof. Herzberg while working together on the subject.

4.1.1 The Concepts of Streams, Channels, and Components

Informally, a component is typically defined as a "physical encapsulation of related services according to a published specification" [8,10]. The underlying concept is the idea of a component which encapsulates a local state or a distributed architecture. Based on this idea, the mathematical notion of a component is rather straightforward: it relates streams of messages (or actions) on input channels to output channels. All required concepts are defined subsequently.

Untimed Streams An *untimed stream* over the set $M = \{m_1, \ldots, m_i\}$, $i \in \mathbb{N}$, is a finite or infinite sequence of elements from M. The elements of M represent messages or alternatively actions. For the sake of brevity, we will refer to messages in the following. As an example, let us assume that M consists of two messages $M = \{m_1, m_2\}$. We now construct an infinite number of streams out of M. For example, the sequences $\langle m_1 \rangle$, $\langle m_2, m_1, m_1 \rangle$, and an infinite series of m_2 messages $\langle \ldots, m_2, \ldots \rangle$ represent each one valid streams over M. The *empty stream* is denoted by $\langle\rangle$.

The set of all finite streams which can be formed over the set M is denoted by M^*. M^∞ denotes the set of all infinite streams over M. Putting the sets M^* and M^∞ together results in the set M^ω of all streams over M:

$$M^\omega =^{\text{def}} M^* \cup M^\infty$$

Timed Streams To introduce the notion of time, we suppose a discrete time model with time intervals of equal length. By adding so-called *time ticks* (represented by a $\surd$ symbol) in a stream, we indicate the progress of time and have introduced the concept of a *timed stream*. For example, the transmission of m_1 in the first time interval prior to the first tick, and the transmission of m_2, m_1, and m_2 (in exactly that order) in the next time interval between the first and the second tick, is noted as the timed stream $\langle m_1, \surd, m_2, m_1, m_2, \surd \rangle$.

The set of all finite timed streams over M is denoted by $M^{\underline{*}}$. That means, there is a finite number of ticks in each individual stream. $M^{\underline{\infty}}$ denotes the set of all infinite timed streams over M with an infinite number of ticks. The conjunction defines the whole set of all finite and infinite timed streams over M:

$$M^{\underline{\omega}} =^{\text{def}} M^{\underline{*}} \cup M^{\underline{\infty}}$$

Note that a stream represents a history of communication, while M^*, $M^{\underline{*}}$ etc. represent a set of communication histories. Whenever we want to express that a stream s represents a concrete instance of a timed infinite communication history, we write $s \in M^{\underline{\infty}}$. As we will see, the notion of communication histories allows us to describe the specification of a component in a very compact manner.

From a mathematical point of view, streams are functions mapping natural numbers to messages.

For instance, stream s_i (starting with message m_1 followed by messages m_2, and again m_2) is uniquely characterized by the function

$$s_i \in \{1, 2, 3\} \rightarrow \{m_1, m_2\}, \text{where } s_i(1) = m_1, s_i(2) = m_2, \text{and } s_i(3) = m_2.$$

Channels A (directed) *channel* is an abstract concept of a communication media that transfers messages from a sender to a receiver; the transfer is uni-directional, immediate, faultless, and the order of messages is preserved. A channel consists of a channel name (also called *identifier*) i and an associated channel type T. The relation identifier/type is usually denoted as $i : T$. In general, a type T is given by a set of messages, which can be potentially sent over the channel. For example, the set $\mathbb{B}\text{it} =_{\text{def}} \{0, 1\}$ may stand for type T of the channel identifier i. Thus, $\text{type}(i) = \mathbb{B}\text{it}$. Formally, the type assignment is given by the mapping

$$\text{type} : i \longrightarrow T$$

The timed stream s, $s \in \mathbb{B}\text{it}^{\underline{\infty}}$, is a valid history of the space of all infinite "bit-ed" communication histories transferable over the channel i.

Channel Valuations Specifically, the messages of a stream s associated with a channel identifier i have to be elements of the channel type $\text{type}(i)$, $i : \text{type}(i)$, meaning the *channel valuation* has to be fulfilled:

$$\forall j \in \{1, \ldots, \#\overline{s}\} : \overline{s}.j \in \text{type}(i)$$

Here, $\overline{s}$ stands for a stream with all ticks removed from s, $\#s$ is the length of s, and $s.j$ returns the jth element of s. The type information can be expressed for an infinite timed stream s by

$$s \in \text{type}(i)^{\underline{\infty}}$$

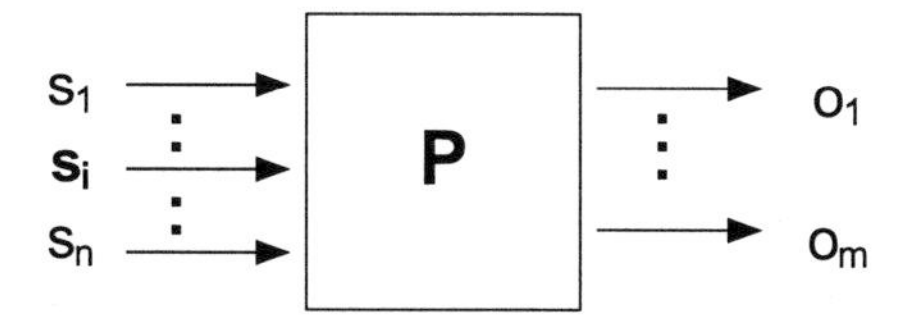

Figure 4.1: Component $\mathcal{P}$ with input and output named channels

Components Seen from a black box perspective, a *component* can be specified by (a) its syntactic interface and (b) its behavior which can be observed at the interface.

The syntactic interface is given by a list of channel names, which we refer to as input/output identifiers (I/O identifiers), and a list of associated channel types. For any component specification S, by $i_S = (i_1, \ldots, i_n)$ and $o_S = (o_1, \ldots, o_m)$ $(n, m \in \mathbb{N})$ we denote its list of input and output identifiers, respectively. $I_S = (I_1, \ldots, I_n)$ and $O_S = (O_1, \ldots, O_m)$ are the corresponding lists of their types. For the association channel identifier to channel type, we also write $i_1 : I_1; \ \ldots; \ i_n : I_n$ (or $i_S : I_S$ for short) and $o_1 : O_1; \ \ldots; \ o_m : O_m$ (or alternatively $o_S : O_S$). We refer to $(i_S : I_S \rhd o_S : O_S)$ as the *syntactic interface* of S; $(I \rhd O)$ is a shorthand notation for that. To state that S is a specification with the syntactic interface $(I \rhd O)$ we write

$$\mathrm{S} \in (I \rhd O)$$

In simpler terms, a component is connected to its environment exclusively by its channels. The syntactic interface indicates which types of messages can be exchanged but it tells nothing in particular about the interface behavior.

For the behavioral description of the component S, we associate type consistent streams to each of the channel names. In favor of a compact notation, the I/O identifiers also designate their respective streams, which represent communication histories; i.e., $i_S \in I_S^{\infty}$ (abbreviated for $i_1 \in I_1^{\infty}; \ \ldots; \ i_n \in I_n^{\infty}$) and $o_S \in O_S^{\infty}$ (for $o_1 \in O_1^{\infty}; \ \ldots; \ o_m \in O_m^{\infty}$) in case of infinite timed streams. The black box behavior of a component is described by predicates characterizing a subset out of all possible I/O histories

$$\mathcal{R}_S \subseteq I_S^{\infty} \times O_S^{\infty}$$

With B_S representing the body of the specification S, this set of I/O histories forms a relation $\mathcal{R}_S$ called the *I/O behavior* of S. Its meaning is defined by the formula

$$(i_S, o_S) \in \mathcal{R}_S \Leftrightarrow B_S$$

The definition of the I/O behavior $\mathcal{R}_S$ is general enough to include non deterministic system behavior as well. An input history may relate to several output histories. $\mathcal{R}_S$ can be characterized in the following styles: equational, assumption/guarantee (A/G), graphical or relational. For instance, we use the graphical style with state machines in our composition filters model. For causality restrictions and more on the specification styles [11] can be consulted.

Definition 7 (Stream-processing function) *A function*
f : $\textbf{\textit{Stream}}_I \rightarrow \textbf{\textit{Stream}}_O$ *from sets of streams to sets of streams is called a* stream-processing function.

We will relate stream-processing functions, or components, to state machines in the aspect modeling framework in the coming chapter.

4.1.2 Composition with Mutual Feedback and Behavioral Refinement

Some more definitions are needed to explain arrangements of components.

Definition 8 (Denotation) *For any timed specification* S *written in the relational style, we define its denotation, written* $[\![\mathrm{S}]\!]$, *to be the formula*

$$i_{\mathrm{S}} \in I_{\mathrm{S}}^{\underline{\infty}} \wedge o_{\mathrm{S}} \in O_{\mathrm{S}}^{\underline{\infty}} \wedge B_{\mathrm{S}}$$

If some output channels of one component are input channels of another component and vice versa, we speak of *mutual feedback*.

Definition 9 (Composition with Mutual Feedback) *Given two specifications:* $\mathrm{S}_1 \in (I_1 \rhd O_1)$ *and* $\mathrm{S}_2 \in (I_2 \rhd O_2)$ *with disjoint sets of output identifiers* $O_1 \cap O_2 = \{\}$, $l = (I_1 \cap O_2) \cup (I_2 \cap O_1)$ *stands for a subset of channel identifiers of* O_1 *and* O_2, *which are input to* I_2 *and* I_1, *respectively. The same type* L *is assumed for the affected internal channel identifiers in* l. Composition with mutual feedback $\mathrm{S}_1 \otimes \mathrm{S}_2$ *is then defined as*

$$[\![\mathrm{S}_1 \otimes \mathrm{S}_2]\!] =_{\mathrm{def}} \exists\, l \in L^{\underline{\infty}} : [\![\mathrm{S}_1]\!] \wedge [\![\mathrm{S}_2]\!]$$

Composition with feedback is powerful enough to express all kinds of communication connections between components provided the channels of the components are consistently named. We illustrate mutual feedback component composition in Fig. 4.2. An example follows.

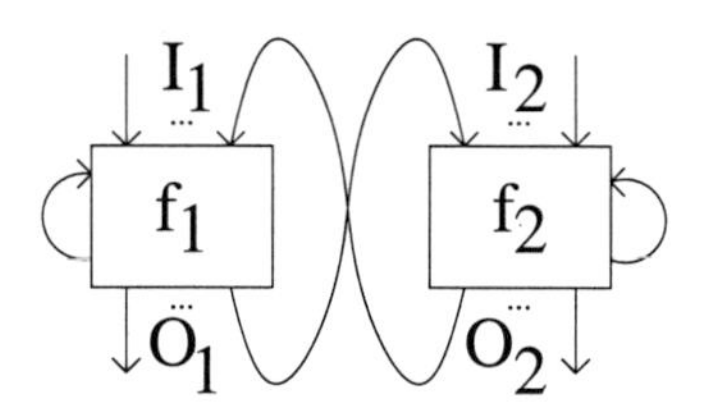

Figure 4.2: Composition of stream-processing functions with mutual feedback

Example If f : $\mathbf{Stream}_{I_1} \rightarrow \mathbf{Stream}_{O_1}$, $f(s) =^{def} \{1.s, 2.s\}$, is the stream-processing function with input channel I_1 and output channel O_1 that outputs the

input stream s prefixed with either 1 or 2, and
$g : \mathbf{Stream}_{I_2} \rightarrow \mathbf{Stream}_{O_2}$, $g(s) =^{def} \{0.s\}$, the function with input (resp. output) channel I_2 (resp. O_2) that outputs the input stream prefixed with 0, then the composition $f \otimes g : \mathbf{Stream}_{I_1} \rightarrow \mathbf{Stream}_{O_2}$, $f \otimes g(s) = \{0.1.s, 0.2.s\}$, outputs the input stream prefixed with either of the 2-element streams 0.1 or 0.2.

Behavioral refinement supports adding properties step by step to a specification, while it is guaranteed that any I/O history of the refined specification, the more concrete specification, is also an I/O history of the given specification; the more abstract one.

Definition 10 (Behavioral Refinement) *Let* S_1 *and* S_2 *be specifications with the same syntactic interface. The relation* $\rightsquigarrow$ *of behavioral refinement is defined by the equivalence*

$$(\mathrm{S}_1 \rightsquigarrow \mathrm{S}_2) \Leftrightarrow ([\![\mathrm{S}_2]\!] \Rightarrow [\![\mathrm{S}_1]\!])$$

Composition with mutual feedback is modular with respect to behavioral refinement. This means that a refinement of components results in a refinement of the system.

4.2 Aspect Weaving

Weaving is a central notion in aspect orientation. This notion may be examined as a refinement relationship of the communication relations between the components of a given model. In a software system, communication is usually regarded as ideal, defect free and almost instant. A channel, as defined above, is such an ideal entity. But at the same time, to describe distribution and other means of communication (e.g. connectionless communication) as well as aspect weaving, the explicit introduction of a communication service generalizes our considerations.

In order to explain weaving in AO as a refinement relationship over the communication channels we recalled the definitions of the explicit means of communication and communication refinement from Herzberg [34] and Herzberg et. al [35]. The author discussed these ideas with Prof. Herzberg and Prof. Broy while working together on a paper on the formal foundations for aspect orientation. The focus of that work is relating AO to a formal theory. The paper was submitted for publication and contains an interesting classification schema that we do not include in this work. The schema will be submitted for publication separately.

4.2.1 Means of Communication

We need to explicitly highlight the means of communication in form of a *communication service*. A communication service might simply provide a static connection-oriented service, thereby reflecting an ideal or non ideal connector. On the other hand, the communication service might also provide a connectionless means of communication, representing for example a message router. A communication service connects only a subset of the input and output channels of one component with those of another one.

Definition 11 (Communication Service) *Let* $\mathrm{S}_1 \in (I_1 \rhd O_1)$ *and* $\mathrm{S}_2 \in (I_2 \rhd O_2)$ *be given components. A* communication service C *for* S_1 *and* S_2 *is a component in* $(I_1' \cup I_2' \rhd O_1' \cup O_2')$ *where* $I_1' \subseteq I_1$, $I_2' \subseteq I_2$, $O_1' \subseteq O_1$, $O_2' \subseteq O_2$. *Its composition is defined by*

$$\exists\ I_1', I_2', O_1', O_2' : [\![\mathrm{S}_1]\!] \wedge [\![\mathrm{C}]\!] \wedge [\![\mathrm{S}_2]\!]$$

and we then write

$$\mathrm{S}_1 \leftarrow \mathrm{C} \rightarrow \mathrm{S}_2$$

We illustrate Definition 11 in Fig. 4.3.

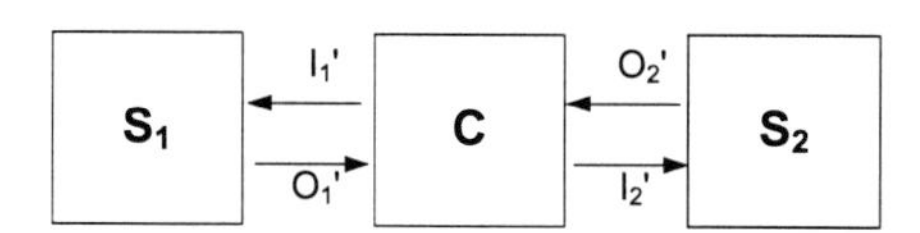

Figure 4.3: Communication service

For simplicity's sake, we assume here and in the following, that all channels are named in a way such that there are no name conflicts. This form of inserting a communication service between two components is basically a generalization of composition with mutual feedback. The notation presented is just a way to semantically highlight that C represents the communication means and is not supposed to be a "usual" component whereas a usual component has a behavior specification, e.g., indicated by the body of the specification as introduced in the paragraphs before Definition 7.

4.2.2 Communication Refinement

Communication refinement relates two specifications of communication services C_1 and C_2 written at different levels of abstraction. Two other specifications, R and A, adapt the interfaces of the communication services and mediate between the abstract specification C_1 and the more concrete specification C_2. R is called the *representation* whereas A is called the *abstraction*.

Definition 12 (Communication Refinement) *Let* C_1, C_2, R, *and* A *be specifications such that*

$$\mathrm{C}_1 \in (\mathrm{I}_1 \rhd \mathrm{O}_1) \wedge \mathrm{C}_2 \in (\mathrm{I}_2 \rhd \mathrm{O}_2) \wedge \mathrm{R} \in (\mathrm{I}_1 \rhd \mathrm{I}_2) \wedge \mathrm{A} \in (\mathrm{O}_2 \rhd \mathrm{O}_1)$$

The relation of communication refinement from C_1 *to* C_2 *is defined as*

$$\mathrm{C}_1 \leadsto (\mathrm{R} \otimes \mathrm{C}_2 \otimes \mathrm{A})$$

There is no need to limit communication refinement to communication services; in fact, the definition of communication refinement is valid for any kind of specification. Nevertheless, in this context, the focus on communication services as the targets of refinement is determined by our subject of study. That is why we also write

$$\mathrm{C}_1 \leadsto (\mathrm{R} \leftarrow \mathrm{C}_2 \rightarrow \mathrm{A})$$

Communication refinement is a special form of behavioral refinement: one component specification can be substituted by an arrangement of three other components.

4.2.3 Aspect Weaving by Communication Refinement

Now, we have everything in place to provide a formal foundation for weaving in aspect orientation. We do so by defining *aspect weaving*:

Definition 13 (Aspect Weaving) *Let the set of components* F_1, C_1, *and* G_1 *describe a model* Z_1, *with* C_1 *being the communication service:*

$$\mathrm{Z}_1 = (\mathrm{F}_1 \leftarrow \mathrm{C}_1 \rightarrow \mathrm{G}_1)$$

Correspondingly, let F_2, C_2, *and* G_2 *describe a model* Z_2, *with* C_2 *being the communication service:*

$$\mathrm{Z}_2 = (\mathrm{F}_2 \leftarrow \mathrm{C}_2 \rightarrow \mathrm{G}_2)$$

Aspect weaving is defined as applying communication refinement such that

$$\mathrm{C}_1 \leadsto \mathrm{Z}_2$$

resulting in a new composition set

$$\mathrm{F}_1 \leftarrow \mathrm{F}_2 \leftarrow \mathrm{C}_2 \rightarrow \mathrm{G}_2 \rightarrow \mathrm{G}_1$$

The composites $\mathrm{F}_1 \leftarrow \mathrm{F}_2$ *and* $\mathrm{G}_2 \rightarrow \mathrm{G}_1$ *are called* slices.

One interesting part of aspect weaving is the shift in viewpoint. Before aspect weaving, a concern describes an independent, coherent network of communicating entities; one could also say that concerns define a *horizontal view*. After aspect weaving, slices dominate the view. A slice assembles parts of different concerns; a slice is said to *crosscut* concerns. In that sense, a slice "vertically" cuts through two or more concerns, see figure 4.4.

Another consequence of our definition is that there is virtually no limit for weaving aspects. Each new "layer" adds further aspects to a slice. This stacks up an aspect hierarchy within a slice. Usually, aspect layers are not commutative: It makes a difference whether logging is first added to a slice and then encryption or vice versa. Also, aspect hierarchies are not abstraction hierarchies. The top level aspect in a slice is not more abstract or concrete than the aspect at the bottom of a slice.

Please note that the arrangement of added components may result in incorporating further communication channels (for instance in F_2 and G_2 in Definition 13) to other layers in the system. This means that the process of *aspect weaving* incorporates new components which may also include communication channels to components in other layers. This way, our approach is as general as possible, providing maximal flexibility for system composition.

Most importantly, the definition given enables us to classify different kinds of aspects. More about this is introduced in the following section.

An aspect can be introduced into the system either by redirecting the flow of messages accordingly or by embracing a base component with an aspect. We call the former *dynamic aspect weaving* and the latter *static aspect weaving*. We model the latter in the coming chapter.

We consider this proposal as a means for modeling aspects at the software architecture level. Our definition of aspect weaving by communication refinement provides a formal underpinning to explain the concepts of aspect orientation.

4.3 Related Work

Ever since aspect orientation has been around, there have been attempts to formally capture the notion of aspects. Many papers are influenced by AspectJ [51] and consequently aim for a formalism regarding clear box AO. On a coarse grained black box level, we identified a comparable work on the approach taken, see [5]. ATKINSON and KÜHNE base their understanding of aspect orientation on interaction refinement, a more restricted variant of communication refinement. Their view on aspects results in a stratified architecture, which is very similar to our aspect layers yet they do not provide a formal foundation for their approach. That such a formal underpinning could be achieved with communication refine-

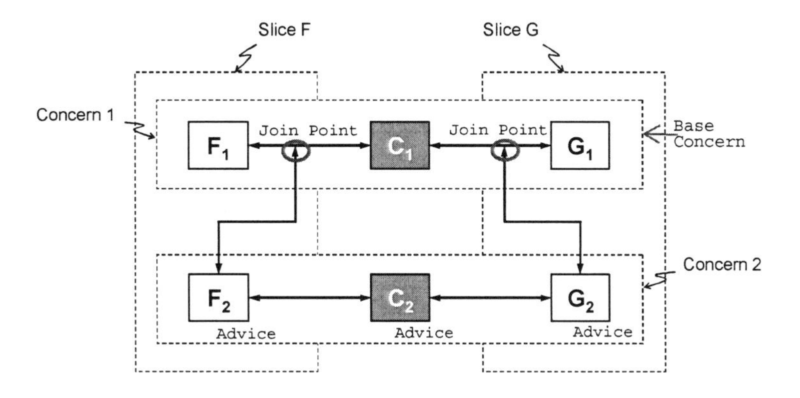

Figure 4.4: Weaving aspects via communication refinement

ment, is first mentioned in [34]. In [35], communication refinement is used for a formal approach to distributed communication systems, which can be considered a special case of aspect orientation.

4.4 Summary

A systematic analysis was given that aspect weaving, a key concept in aspect orientation, can be explained by a composition technique based on communication refinement, which is a special form of behavior refinement. Seen in this way, AO is not really a new concept. It is something that has been used for at least two decades already in the reference models of OSI and TCP/IP.

However, the selection mechanism, e.g., "join point model" does represent, together with the notion of weaving, an advancement with respect to models like refinement layers as illustrated by the OSI/ISO reference model.

Putting this chapter in relationship to the previous ones, we showed how the concepts of AO can be explained at the software architecture level by explaining weaving through communication refinement and by means of our composition model in the coming section. Our approach toward weaving and composition provides the foundation for this theory of aspects. We may use it, for instance, to model aspect interaction.

Chapter 5

Modeling Functional Aspects

Our affections have always an astounding talent to disguise themselves as philosophy of life. -Hermann Hesse [1]

The purpose of this chapter is to exemplify the use of our aspect modeling proposal, i.e., theory with a more practical application. Moreover, we explored this modeling proposal at the hand of security examples, since we considered that a formal foundation for dynamic security aspects in aspect-oriented development is still not provided yet since the time we started working on the subject more related work has appeared. We can neither argue that our proposal is the only one, see the related work section, nor say that this proposal is demonstrated to serve the general case of aspect composition. However, it contributes a formal approach to the problem that is in nature different to other recent proposals. For instance, some formal approaches towards analysis of aspect interaction based on graph transformation and even more if we consider approaches towards aspect interaction that are mostly syntactical.

We introduce a composition model that substantiates the analysis of aspect interaction at the software architecture level. This proposal is not intended for the analysis of aspect orientation at code level yet in order to reason about aspect composition our framework provides syntax and semantics based on the underlying theory FOCUS. These underpinnings may allow for a precise analysis of aspect composition at the architectural level.

The definitions of: streams, components and their composition, introduced in Sect. 4.1 provide the foundation that allows analyzing the history of messages that either enter or leave a component. This is what we call the channel history of a component. Through the channel history of components we may analyze their interaction. For instance, we may observe whether a given message m received as

[1] Unsere Neigungen haben stets eine erstaunliche Begabung, sich als Weltanschauung zu maskieren. -Hermann Hesse

input at channel s_1 in a component $\mathcal{P}$ (Fig. 4.1) at time **t** is present in some output channel o at time **t+1**. Similarly, given that $\mathcal{P}$ can be specified under this theory as a stream processing function (component) with precise syntax and semantics, we may follow its output history to verify whether a given system property[2] is actually fulfilled. That is the reason for the use of this formalism to study aspect interaction.

The particular case study i.e. example is provided in Sect. 5.3. The case study consists of a *secure channel*[3] and an *authentication* aspect. Their composition is performed in Sect. 5.4. Additionally; in Sect. 5.5 we determine under which conditions these two aspects can be securely composed together. In Sect. 5.6 we recall a tool framework for verification of (predefined) security properties in UML models. Following the diagnosis of these tools we propose that the security failures can be corrected by weaving the required aspect using this framework. We further relate our work at the modeling level to aspect weaving at the code level with an example in Java and COMPOSE* in Chap. 5.7. We close this chapter with a comparison to related work and conclusions.

5.1 Components, i.e., Stream Processing Functions Characterized by State Machines

We briefly recall that FOCUS [11] describes a system based on input/output relations on sets of histories of externally observable events. A system is divided into components. The behavior of a component is described by the relationship between its external input and output histories, defined as stream-processing functions (where "streams" are the sequences of input and output values of the system, to be defined below). In this manner, we may obtain a *black box view* of the component in question. This theory also allows us to distinguish between *elementary* and *composite* specifications. Composite specifications are built from elementary specifications using constructs for composition and network description.

We will build composite specifications out of elementary specifications by composing stream-processing functions. Please bear in mind that our CF model considers the communication channel as the entity at which the composition is realized which is characterized in the previous channels as communication refinement. The tables in this section indicate the places, i.e., channels at which the communication service is modified or refined with the CF's.

We choose to represent stream processing function by state machines as ex-

[2]i.e. set of behaviors see [2]

[3]To be more precise this is actually a *secrecy-enforcing channel*. We call it "secure channel" in this work to facilitate reading.

plained in [11]. We take the definitions of state machine and state transition diagram from [9].

State transition relations are described the by state transition rules. These are logically represented with the help of assertions that contain the state attributes *v* as identifiers in a primed form *v'* and in an unprimed form *v* as well. Unprimed identifiers relate to the values of the given attribute in the state before the state transition and the primed identifiers to the values of the attributes in the state after the transition.

Transition rules are of the syntactic form [$\mathcal{P}$] x:e / y:b, where $\mathcal{P}$ is a guard which is a stated assertion referring only to the local attributes of the system. The identifier x denotes an input channel and y an output channel, e and b are messages of respective types of the channel. A transition is fired when an appropriate message is received and the specified guard is fulfilled.

State machines are described by a state assertion U that characterizes the set of initial states and a finite set of state transition rules R of the form presented above.

A state transition specification $\mathcal{S}$=[$\mathcal{I}$/$\mathcal{O}$, **attribute** B; **initial** $\mathcal{U}$;$\mathcal{R}$] consists of a given set of typed input channels $\mathcal{I}$, a given set of typed output channels $\mathcal{O}$ and a set of typed attributes B; furthermore, it contains a state assertion $\mathcal{U}$ which characterizes the set of initial states and a set $\mathcal{R}$ of state transition rules.

State machines can be described by a **state space**, **state transition rules**, and an **assertion on the initial states** of the machine. The state is specified by a set of typed attributes. Each valuation of these attributes describes a state. In the assertion related to the initial state, we referred to the state attributes and the output channels. We modeled the state transitions between the system states and the input and output sequences of messages arriving through the channels. State machines are represented by state transition diagrams.

A state transition diagram is a graph with nodes labeled by control states and arcs labeled by state transitions, where one control state or a set of them is marked as being initial.

In the state transition diagrams we presented here, the channel type is indicated below the graph, together with the corresponding expressions that each may send (respectively receive).

We denote a specification $\mathcal{S}$ as $[\![S]\!]$. So that, for instance, the specification represented as a state machine in Figure 5.2(a) is denoted $[\![Send_E]\!]$. The input and output channels are specified in the lower part of the diagrams.

Now, we have everything in place to introduce the concern composition model.

5.2 Composition Model

The composition model we proposed is inspired by composition filters from [7]. This model is based on message interception over communication channels between objects. Input and output filters are defined around a base object. The filters select messages based on given criteria and either perform a process on the selected messages or send the message (intact or altered) to a predefined recipient. If the message was not selected, it is then forwarded to the next filter. Input filters select incoming messages, output filters select outgoing messages. The CF model we refer to is implemented in the COMPOSE* framework [15]. It is defined at the programming level, not at the software architecture level as is ours. COMPOSE* is one of the first aspect-oriented languages and it is mature enough to be applied industrially. It is important to mention that similar implementations of composition filters have appeared in the form of proxies, portable interceptors and reflection techniques in Jboss yet we did not explore them here. We chose COMPOSE* as inspiration for our composition model.

The CF model exploits the fact that objects in the OOP can only communicate by sending messages. Based on this principle, a set of filters is defined together with an advice. The filters select incoming and outgoing messages according to logical conditions and execute the respective advice. These filters are set syntactically over the signature of objects.

We will introduce an architecture level composition filters model. In the case of a *black box* view, components (defined in Sect. 5.1) communicate only through their input and output channels i.e. interfaces. In the case of a *glass box* (also called *clear box*) view, the communication is analyzed at the level of internal communication channels. The main elements we propose for a formal composition filters model are a set of base components, a set of component filters which are each defined as a stream processing function and a table relating both.

Our composition model relies on the concept of components as interactive system entities. We considered the base concern as a set of interconnected components. This can be understood as a network of components that constitute a software system. The aspects are modeled as a set of filters and are also defined as components. The filters are composed with the base concern by weaving them as components added in the communication channels of selected members of the base set.

We characterize *weaving* as mutual feedback component composition using the composition operator $\otimes$ in Definition 9.

An `Advice` is a (functional) element which augments or constrains other concerns at JP's matched by a pointcut expression, i.e., pointcut delimiter. In our model the advice is specified by state machines expressing the behavior of a given filter.

Composition Filter Given a set of n stream processing functions a Composition Filter (CF) is defined as a set: $CF =^{def} \{CF_1, \ldots, CF_n\}$, where CF_j is a component with index $j \in \mathbb{N}^{\{1,\ldots,n\}}$

The CF model in [7] allows one to direct messages to internal or external objects. We believe that, although powerful, this mechanism can be difficult to understand, to control and to verify whether it is actually being used as intended yet our model also allows to direct messages to components other than the ones originally considered.

In this work, we restricted our analysis to a composition model defined over the external channels of components (at the level of component interfaces), since this is sufficient for our purposes and allows us the kind of analysis described in Sect. 5.5. Since the filters are defined as components, we may provide for channels, i.e., streams from the filter to components in other parts of the system. Therefore, our model can perform message redirection if needed.

In the following, we introduce a way to designate the insertion of the filters expressed in Table 5.1. For that purpose we adapt the aspect abstraction proposed in [24] (for an object-oriented model) to our component-oriented approach. The first part of the table provides the aspect name and the second one its identifier. The third row of the table designates the set of components ($\mathbb{M}$) affected by the aspect as well as the base component around which the filters will be composed. This *base component* is intended to act as a pivotal point around which the aspects are connected. In the case we explored here, the base element is the communication channel in Fig. 5.1. The fourth row of the table relates each component in the filter set to the formal specification of the intended behavior i.e. *advice* (expressed as state machines). The fifth row contains our Pointcut (or PCD), in other words, the composition order.

In Fig. 5.1 the process of weaving the aspects with the channel and other parts of the system is illustrated. The tables are explained in more detail in sections 5.3 and 5.5 at the hand of the particular case study.

The *weaving* process may be considered as a transformation function of the form

$$\Psi : \{Components\} \times CF \mapsto \{Components\}$$

We first explored this idea in terms of model transformation in [26]. Now, in this work, we relate the actual realization of the weaving function to the composition operator (⊗) from Sect. 5.1 together with the specification of the weaving order in the corresponding tables.

We briefly recall that there are two possible weaving techniques from the perspective of *when* the weaving process is performed. On the one hand, dynamic weaving if the process is performed at run time, on the other hand, static weaving if the process is performed during compilation. Our proposal belongs to the second group. It performs static weaving without tool support.

The formal definitions in Sect. 5.1 provide precise syntax and semantics to our model. On this basis we specify a generic secure channel in the next section together for which we will then perform a formal analysis of (security) aspect composition.

The relation of our proposal to object orientation can be illustrated by equating the methods of a class as a subsystem on its own. We recall that the composition filters approach of [7] stresses the fact that the communication between methods in OOP is abstracted into the messages sent between objects. We model this message sending as communication channels between components.

5.3 Specifying Security Aspects

We now will illustrate the proposed composition model on the case of a web banking system taken from a joint project with a major German bank. We considered the base concern to be a Client and a Web server communicating over a given Internet medium (also referred to as communication channel in the tables). The aspects are a secure channel as defined in [43] and an authentication protocol from [29].

The web banking system consists of an Internet based application that allows clients to complete and sign a digital order form. The main security concerns of this application we considered for the analysis are two. First, user data must be kept confidential. This implies the use of a secure communication channel or a protocol that ensures the privacy of the data. We build in section 5.3.1 our system over a secure channel that may be considered generic and usable for similar problems. This secure channel has been modeled against a generic attacker and fulfills the property of ensuring user's privacy. Secondly, it is required that orders may not be submitted in the name of other users. On account of this when the user logs-in an authentication protocol runs and a confidential connection is established. The authentication protocol is based on SSL and it constitutes the second concern that we will compose with the secure channel. This demonstrates the application of our approach.

An overview of the system is shown in Fig. 5.1. This figure helps to illustrate, as we mentioned before, that our CF model considers the communication channel,

Aspect Name
Aspect identifier
Set of base components $\mathbb{M} = \{C_0, ..., C_K, ..., C_m\}$ Base Component: $Base = C_K$
Composition Filter: $[\![\mathsf{CF}_1]\!] = [\![\mathit{Specification}_1]\!]$, $[\![\mathsf{CF}_2]\!] = [\![\mathit{Specification}_2]\!]$ $\ldots$ $[\![\mathsf{CF}_n]\!] = [\![\mathit{Specification}_n]\!]$
Weaving order of CF to base concern (Pointcut Designator): $[\![C_0]\!] \otimes \cdots \otimes [\![\mathsf{CF}_1]\!] \otimes [\![Base]\!] \otimes [\![\mathsf{CF}_2]\!] \otimes \cdots \otimes [\![C_m]\!]$

Table 5.1: Aspect weaving to base concern: Generic case

in this case the place at which the secure channel is incorporated, as the entity at which the composition is realized. This is explained in the previous chapter as communication refinement. The tables in this section indicate the places, i.e., channels at which the communication service is modified with the CF's.

We focus on the examination of the two security concerns and their interaction. Nevertheless, the approach is of interest for related problems. For instance, composing a different authentication protocol over the secure channel, or modifying the secure channel while preserving the outer layers of the system and analyzing whether the security concerns still hold. As shown in Fig. 5.1 we build the authentication process over the secure channel. Both aspects are first treated independently of one another and are afterwards woven together with the base components using the composition operator $\otimes$ defined in Sect. 5.1. From the software engineering point of view, keeping changes as modular as possible is of relevance and even more if we need to preserve complex concerns as the ones related to security. What we aim at is allowing software developers to make use of a verified set of components. Constructing a secure channel that guarantees `secrecy` and a protocol that guarantees `authenticity` each as an aspect allows one to later reuse both in other parts of the system. The former can be found in Sect. 5.3.1, the latter in Sect. 5.3.2. Nevertheless, our model may not be useful in case of security properties that have to be ensured through system level support, e.g., through real time filters.

5.3.1 A Secure Channel Aspect

We define the secure channel aspect as the composition filter specifications of a sender (Send_E) and a receiver ($\mathrm{Receive}_E$). Both are weaved with the authentication aspect in Sect. 5.4. We then analyze the composition of the two aspects in a

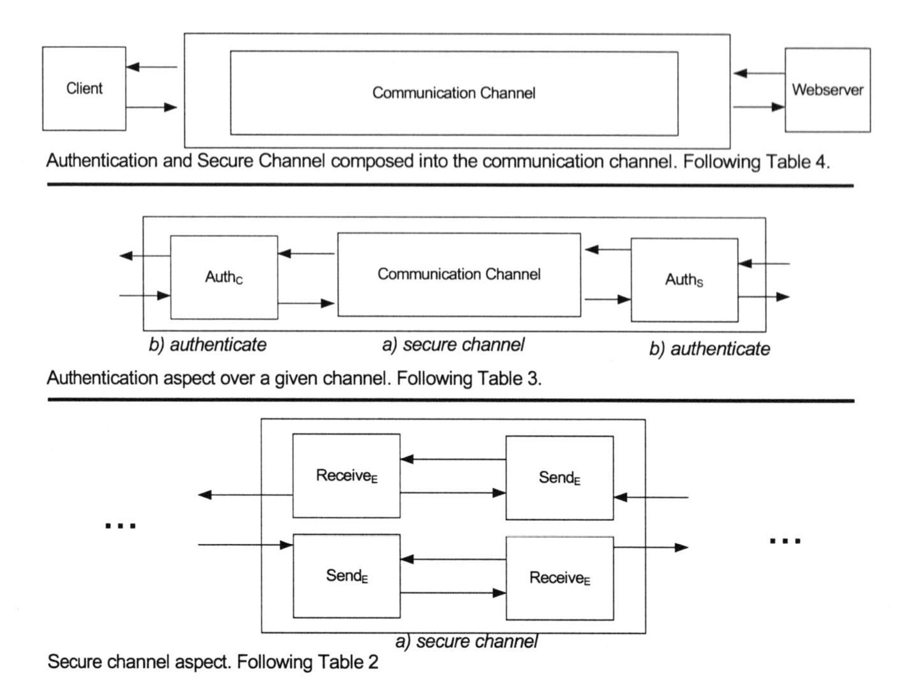

Figure 5.1: Architectural view of the process. Weaving as component composition

general setting in Sect. 5.5.

The secure channel aspect consists of two filters. On the one hand, the sender filter that will be coupled to the client and the communication channel. The sender is in charge of encrypting and signing the messages that will be further sent over the communication channel. On the other hand, we have the receiver filter that gets the messages from the channel, decrypts and unsigns the message, and sends it further.

The next two specifications represent the security aspect that we will compose on a base concern. We delineate the weaving of the secure channel over the communication channel in Table 5.2.

The specification in Fig. 5.2(a) represents the sender side of our first aspect according to the above security property. The state machine represents the specification. The sender retrieves the signed and encrypted symmetric session key k_j from the receiver, checks the signature and encrypts the data under the symmetric key. Encryption is performed together with a sequence number c to avoid replay.

The secrecy property considered here relies on the idea that a system specification preserves the secrecy of a piece of data d if the system never sends out any

Aspect Name: Secure channel
Aspect identifier:CF_{SC}
Set of base components $\mathbb{M} = \{Communication\ channel\}$ Base Component: $Base = Communication\ channel$
Composition Filter: $[\![\mathsf{CF}_1]\!] = [\![Send_E]\!]$, $[\![\mathsf{CF}_2]\!] = [\![Receive_E]\!]$
weaving order of CF to base concern: $[\![\mathsf{CF}_1]\!] \otimes [\![Base]\!] \otimes [\![\mathsf{CF}_2]\!]$

Table 5.2: Aspect weaving to base concern: Secure channel

information from which *d* could be derived, even in interaction with an adversary.

The second part of the secure channel aspect, namely the receiver, is represented in Fig. 5.2(b). The receiver first gives out the key k_j with a signature and also with a sequence number j, and later decrypts the received data checking the sequence number.

5.3.2 An Authentication Aspect

In this section, we will specify the authentication aspect. Its composition with the secure channel from the previous on is examined in Sect. 5.5.

We explain a typical run of the authentication aspect, specified as state machines in Fig. 5.3(a) and Fig. 5.3(b). The client sends the authentication client a *clientHello* message which is forwarded to the server. Afterwards, a randomly generated number (nonce) is sent by the web server. The client signs this nonce with his private key K_C and sends it back together with a global identification number (GID) which is signed using the Key provided by a certification authority K_{CA}. The server checks the signature, the certificate and the GID and sends the client a Data Form. In the specification $Auth_S$ we do not establish a communication channel with the web server itself – for simplicity we assume the data form is generated by the web server and provided to the authentication server.

We may also define other aspects similarly. Once the base component(s) and aspects are specified we proceed to define how they are actually related as a system in the coming chapter. This is actually performed by relating the channel names. This defines the weaving ordering and the resulting system. Therefore, channel renaming is a necessary step for weaving. Just as in AOP the underlying framework ultimately relates aspects and base code at the level of names in class methods or data, as is the case in the prevailing aspect languages.

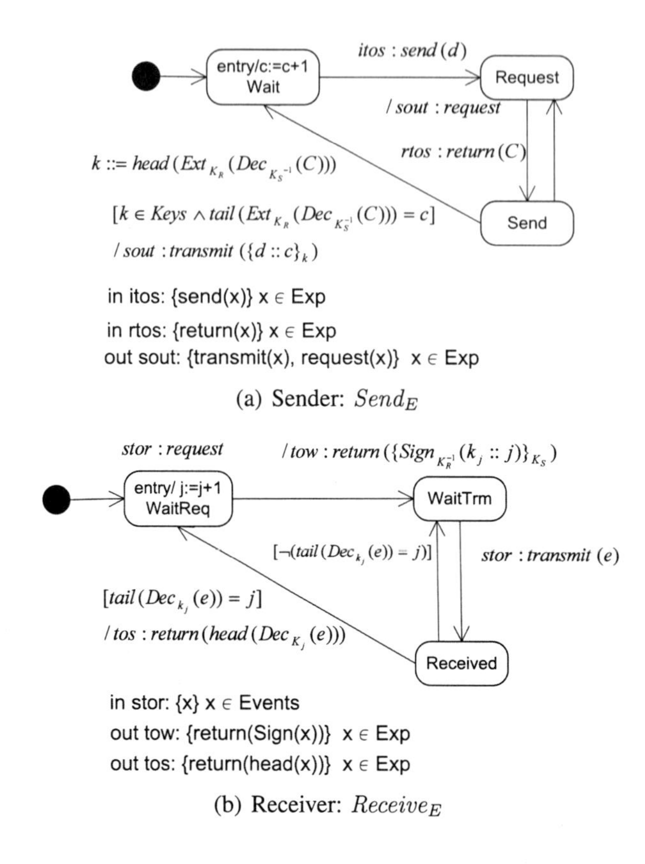

(a) Sender: $Send_E$

(b) Receiver: $Receive_E$

Figure 5.2: State transition diagram secure channel

5.4 Framework I. Composing Security Aspects

The authentication aspect is composed over the secure channel aspect which is first established and generates the session keys. The orders will be signed based on these keys and therefore allow us to guarantee that the bank orders cannot be sent in the name of other users which is the second aspect we need to consider.

Similarly to the secure channel aspect, the authentication aspect is implemented as filters composed with the client and the server. Table 5.3 establishes how this aspect is composed over the communication channel. As designated in Fig. 5.1, the weaving of the CF's to the base concern is:

$$[\![Client]\!] \otimes [\![Auth_C]\!] \otimes [\![CommunicationChannel]\!] \otimes [\![Auth_S]\!] \otimes [\![Webserver]\!]$$

$$\text{Where } [\![CommunicationChannel]\!] = [\![Send_E]\!] \otimes [\![Receive_E]\!]$$

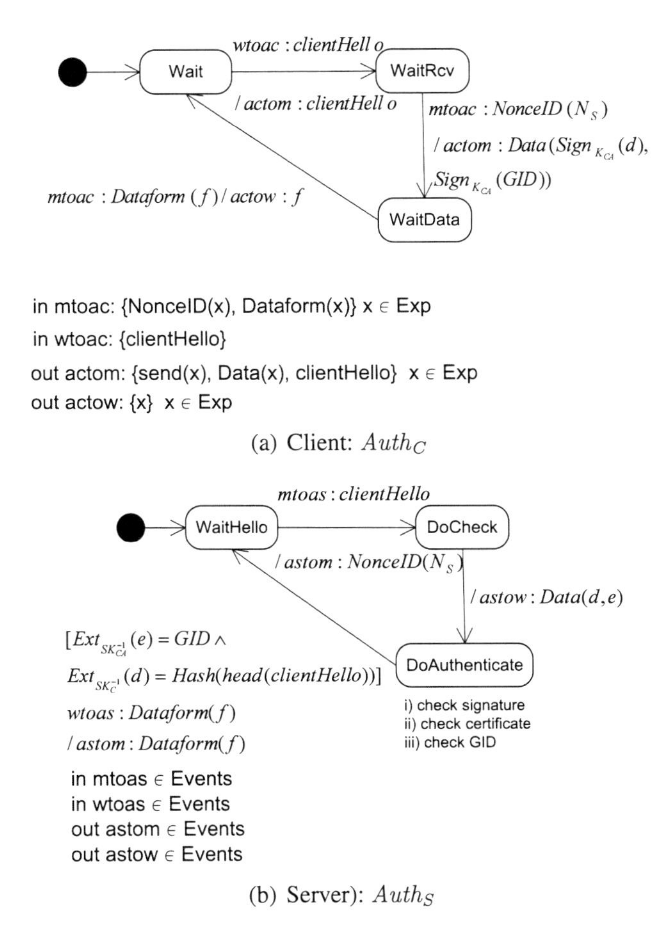

(a) Client: $Auth_C$

(b) Server): $Auth_S$

Figure 5.3: State transition diagram authentication

In order to send and receive through the secure channel, $Send_E$ is connected on the communication channel from the `Client` to the `Server` with the client side $Auth_C$ of the authentication protocol, and when the message is sent from the `Server` to the `Client`, $Send_E$ connects to $Auth_S$. This is illustrated in Fig. 5.1. Ultimately, the channel names are related as indicated below[4]:

I. Communication Channel:

I. $[\![Send_E]\!]$.sout = $[\![Receive_E]\!]$.stor

II. $[\![Receive_E]\!]$.tos = $[\![Send_E]\!]$.rtos

[4]Notation: $[\![SpecificationName]\!]$.Channel name

II. Authentication and Communication Channel

I. $[\![Auth_C]\!]$.actom = $[\![Base]\!]$.itos

II. $[\![Auth_C]\!]$.mtoac = $[\![Base]\!]$.tow

III. $[\![Auth_S]\!]$.mtoas = $[\![Base]\!]$.tow

IV. $[\![Auth_S]\!]$.astom = $[\![Base]\!]$.itos

Aspect Name: Authentication
Aspect identifier:$\mathsf{CF}_{Authentication}$
Set of base components $\mathbb{M} = \{Communication\ channel\}$ Base Component: $Base = Communication\ channel$
Composition Filter: $[\![\mathsf{CF}_1]\!] = [\![Auth_C]\!]$, $[\![\mathsf{CF}_2]\!] = [\![Auth_S]\!]$, $\mathsf{CF}_{Authentication} = \{[\![\mathsf{CF}_1]\!], [\![\mathsf{CF}_2]\!]\}$
Weaving order of CF to base concern: $[\![\mathsf{CF}_1]\!] \otimes [\![Base]\!] \otimes [\![\mathsf{CF}_2]\!]$

Table 5.3: Aspect weaving to base concern: Authentication

Aspect Name: Authentication and Secure Channel
Aspect identifier:CF_{AuthSC}
Set of base components $\mathbb{M} = \{Client, Webserver, \mathsf{CF}_{SC}\}$ Base Component: $Base = \mathsf{CF}_{SC}$
Composition Filter: $[\![\mathsf{CF}_1]\!] = [\![Auth_C]\!]$, $[\![\mathsf{CF}_2]\!] = [\![Auth_S]\!]$
Weaving order of CF to base concern: $[\![\mathsf{Client}]\!] \otimes [\![\mathsf{CF}_1]\!] \otimes [\![Base]\!] \otimes [\![\mathsf{CF}_2]\!] \otimes [\![\mathsf{Webserver}]\!]$

Table 5.4: Aspect weaving to base concern: Authentication and secure channel

Altogether, in Sect. 5.3 we introduced the two aspect specifications that represent functional *aspects*. First, the secure channel's specification in Sect. 5.3.1. Second, the authentication aspect in Sect. 5.3.2. The next step is to weave them in a way that allows us to analyze whether the composition of both specifications respects and entails the desired security properties, namely *secrecy* and *authentication*. Weaving is performed according to the scheme in Sect. 5.2. It is a case of static weaving yet the framework can be extended to consider dynamic weaving, since once the components are verified at the phase at which they are composed to the system does not affect its behavior (at least the way components and therefore aspects are defined in the formal theory recalled in Sect. 5.1).

5.5 Framework II. Analyzing the Composition of Security Aspects

Following from the previous chapters, in which we defined our aspects and set their weaving, now, we may analyze the *secure channel* and *authentication* aspects. The idea is to determine under which conditions they can be securely composed together. The authentication aspect is now implemented over the secure channel as shown in Table 5.4.

The resulting system is specified:
$[\![Client]\!] \otimes [\![Auth_C]\!] \otimes [\![SecureChannel]\!] \otimes [\![Auth_S]\!] \otimes [\![Webserver]\!]$, where
$[\![SecureChannel]\!] = [\![Send_E]\!] \otimes [\![Receive_E]\!]$

We define how this aspect is woven over a given communication channel in Table 5.3.

Specifically, the method we applied here can be summarized in three steps. First, we established that the secure channel defined in Sect. 5.3.1 is generic in the sense that it can securely be composed with a system that satisfies certain saneness condition[5]. Then, we established the properties that the resulting system should preserve. The crpytographic model is the Dolev-Yao cryptographic model. The theorem and proof are recalled from [27].

Theorem 1 *The secure channel aspect preserves the secrecy of the variable* d *from adversaries whose knowledge before initialization of the system does not include any values in the set* $\{\mathsf{K}_\mathsf{S}^{-1}, \mathsf{K}_\mathsf{R}^{-1}\} \cup \{\mathsf{k}_n, \{x :: n\}_{\mathsf{k}_n}\}$ *and includes only such values of the form* $Sign_{\mathsf{K}_\mathsf{R}^{-1}}(\mathsf{k}' :: m)$ *for which we have* $\mathsf{k}' = \mathsf{k}_m$ *for all* $m \in \mathbb{N}$ *and* $\mathsf{k}' \in$ ***Exp***.

[5]for example, that the system itself does not send the secret values to the adversary outside the secure channel

Proof

The proof is of informal nature.
Note that the adversary knowledge set $\mathcal{K}_A$ is contained in the algebra generated by $\mathcal{K}_A^0 \cup \{\{Sign_{\mathsf{K}_R^{-1}}(\mathsf{k_i} :: \mathsf{j})\}_{\mathsf{K_S}}\}$ and the expressions $\{\mathsf{d} :: n\}_{\mathsf{K}}$ for inputs d, where $\mathcal{K}_A^0$ is the initial knowledge of the adversary: Firstly, the adversary can obtain no certificate $\{\{Sign_{\mathsf{K}_R^{-1}}(k :: \mathsf{j})\}_{\mathsf{K_S}}\}$ for $k \neq \mathsf{k_j}$, because the Receiver object only outputs the certificates $\{Sign_{\mathsf{K}_R^{-1}}(\mathsf{k_j} :: \mathsf{j})\}_{\mathsf{K_S}}$ (for $\mathsf{j} \in \mathbb{N}$) to the Internet. Secondly, the sender outputs only messages of the form $\{\mathsf{d} :: n\}_k$ to the Internet, for inputs d and any $k \in$ **Keys** for which a certificate $\{Sign_{\mathsf{K}_R^{-1}}(k :: n)\}_{\mathsf{K_S}}$ has been received. Here k must be K_n since no other certificate can be produced (since the key K_R^{-1} is never transmitted). Note also that $\mathcal{K}_A^{\mathsf{P}} = \mathcal{K}_A^0$ since there are no components accessed by the adversary.
Also, the values that an adversary may insert into the Internet link may only delay the behavior of the two objects regarding $\mathsf{outQu}_{C'}$ since the adversary has no other certificate signed with K_R^{-1} and does not have access to the key K_R^{-1}, and because of the transaction numbers used. Thus any other value inserted is ignored by the two objects.

This means that in particular that the secure channel aspect can be securely composed with any aspect which follows the saneness conditions required in the above result. This gives us a general result on aspect composition, instantiated at the case of the secure channel aspect.

The protocol that implements the authentication aspect from Sect. 5.3.2 has indeed been verified, not only to provide the authentication aspect as hoped, but moreover to be secure in the sense of the assumptions in the theorem above, using a model checker in [29]. Thus we can actually apply the theorem here. In particular, we can compose both aspects and obtain a composed *secure authentication* aspect. Applying this aspect, again given the saneness assumptions of the theorem, now results into a system which provides both secrecy and authentication.

Proposed methodology for verification of security aspects

I. Specify the base component(s) and aspects as state machines

II. Define the composition order and the composition tables

III. Relate the state machines renaming their channels accordingly

IV. Based on the specification of each aspect and the property to be explored, defining a theorem

V. Based on the channel histories find a proof for the theorem.

5.6 Model Verification Framework

Please note that the framework was neither created as part of this thesis work nor was specifically conceived for aspect orientation. We recall it to illustrate that modeling of aspects with the FOCUS formalism as proposed in this thesis, allows relying on formal verification and sound methods for software engineering. This verification framework is used to demonstrate the practical application of our aspect modeling approach. We can also portray it as evidence for the methodical benefit of AO. Since we may verify isolated pieces of software, in this case a secure channel and an authentication protocol, and then rely on its secure properties to remedy failures in other parts of the model or code. We suggest that this proposal to AO can be generalized.

To explain the use of verification tools together with our composition model, consider a UML model annotated with some stereotypes. These stereotypes represent security properties that the model should fulfill at the points in which the stereotypes are present. Say for instance, a link marked «`no down-flow`» means that the information should not be seen by users with lower access rights. Such properties are formulated in UMLsec which is a *lightweight* extension to the UML, and constitutes a flexible framework to define dynamic and static properties in UML diagrams. UMLsec specifies important security properties such as the stereotype «`high`» that denotes dependencies that are supposed to provide the respective security requirement for the data that is sent along them as arguments or return values of operations or signals. The stereotype «`encrypted`» in a deployment diagram denotes the kind of communication link and the associated threats in view of a default or insider attacker. The stereotypes are explained in depth in [43]. In Fig. 5.7, we have a dependency between `web` and `bank` tagged with both stereotypes. The UML model of the Web Store was checked against internal and external attackers particularly in view of the stereotype «`high`». It followed that the link between web and bank were prone to attacks, we focused on this particular result. Therefore, an encryption mechanism to make the channel secure along with an authentication mechanism was needed.

The framework in Fig. 5.4 is used as follows. The developer creates a model and stores it in the UML 1.5 /XMI 1.2 file format. The file is imported by the UML verification framework into the internal MetaData Repository (MDR) repository. MDR is an XMI-specific data-binding library which directly provides a represen-

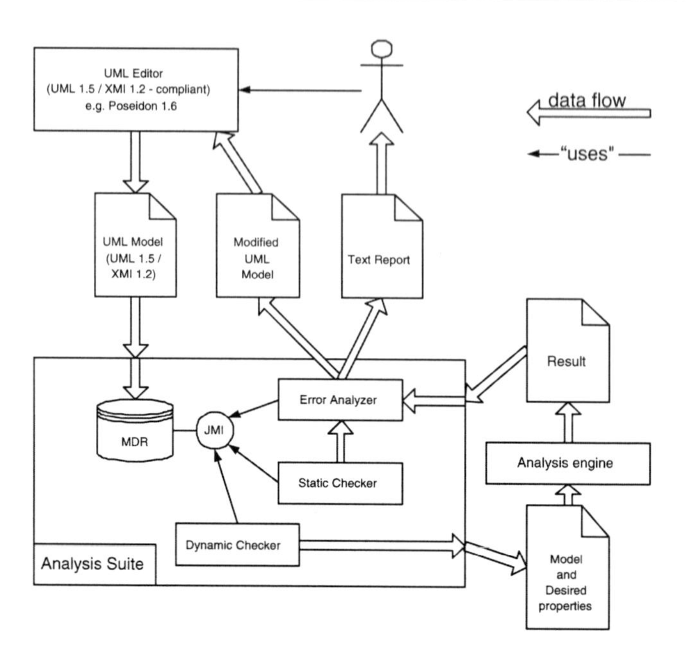

Figure 5.4: UML verification framework from [45]

tation of an XMI file on the abstraction level of a UML model through Java interfaces (JMI). This allows the developer to operate directly with UML concepts, such as classes, statecharts, and stereotypes. It is part of the Netbeans project [61]. Each plug in accesses the model through the JMI interfaces generated by the MDR library, they may then receive additional textual input and they may return both a UML model and textual output. The two exemplary analysis plug ins proceed as follows: The static checker parses the model, verifies its static features, and delivers the results to the error analyzer. The dynamic checker translates the relevant fragments of the UML model into the automated theorem prover input language. The automated theorem prover is spawned by the UML framework as an external process; its results are delivered back to the error analyzer. The error analyzer uses the information received from the static checker and dynamic checker to produce a text report for the developer describing the problems found, and a modified UML model, where the errors found are visualized. Besides the automated theorem prover binding presented in [45] there are other analysis plug ins including a model-checker binding [46] and plug ins for simulation and test sequence generation.

The framework is designed to be extensible: advanced users can define stereotypes, tags, and first-order logic constraints which are then automatically translated to the automated theorem prover for verification on a given UML model.

Similarly, new adversary models can be defined and integrated into the tool suite.

The framework and its application on secure software development have also been introduced in [44, 45]. The user web interface and the source code of the verification framework is accessible at [75].

5.7 Reusing Aspects

In Sect. 5.3, we introduced our two security aspects on the case of a banking system. The idea was to discuss on the one hand the importance of these security protocols, and on the other to abstract from the problem a generic secure channel and an authentication aspect. In this part, we use these two aspects on the case of a web store design. This shows how our method allows for the definition and verification of security aspects that may later be (re)used as building blocks on existing software models, such as the web store whose deployment diagram is shown in Fig. 5.7.

The UML model of the Web Store was checked using the tools introduced in Sect. 5.6. It was checked against internal and external attackers. The analysis with the UMLsec tool has shown that the link between `web` and `bank` was prone to insider attacks, given this particular result an encryption mechanism to make the channel secure is needed.

A class diagram of the Web Store is shown in Fig. 5.6. In order to illustrate the use of our specifications as aspects, we relate the State Machines $Auth_C$ and $Auth_S$ to the *join points* in the message flow of Fig. 5.7. Namely, the communication points where a transaction between the bank and the web is realized. Moreover, we relate $Send_E$, $Receive_E$, to the message calling of $Auth_C$, and $Auth_S$ and view it as a *Shared Join Point*, meaning a Join Point at which two or more aspects are interacting. We illustrated the use of these specifications as aspects by relating them to an implementation of the web store design using COMPOSE* as code level aspect weaver. We would like to emphasize that the proposal is to allow aspect weaving at two different levels. On the one hand, at the modeling level, on the other at code level.

We further explore weaving at the modeling level (upper part of Fig. 5.5) and suggest that code be generated from the resulting model. We might also, specify the individual aspects, the base components, and generate code from these. After that, weave them with a code level weaver (lower part ofFig. 5.5) and obtain the resulting system at code level. The expected result should be equivalent, though this was out of the scope of this work.

A sample code of the *Dress4Less* web store is implemented on J# (.Net), the pointcut designator is defined as the *PaymentSecurity* concern in COMPOSE* and is shown in listing 5.1. This concern specifies two filters, *authenticate_filter* refer-

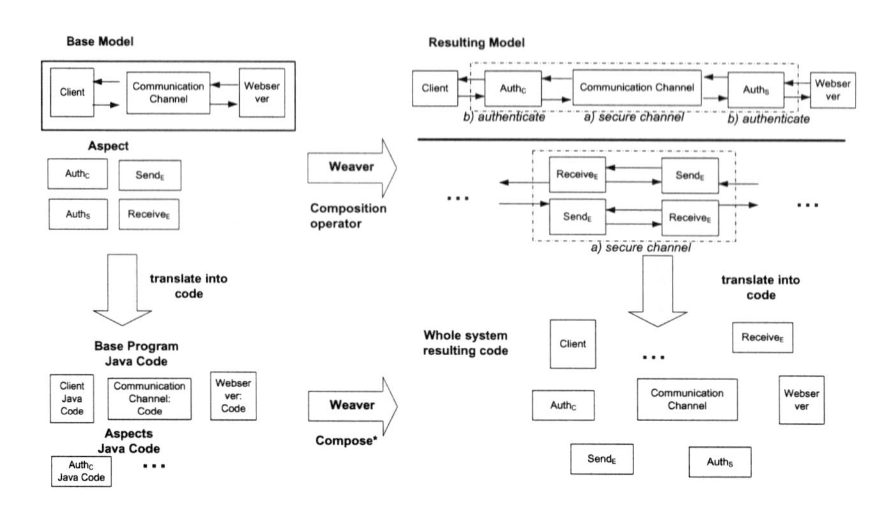

Figure 5.5: Overview of the methodology. Model and code level weaving

ring to the authentication aspect, and *encrypt_filter*. In the *superimposition* section of the concern (line 6 in listing 5.1) we define the methods of the classes *Bank* and *Web* from Fig. 5.6 into which the security aspects are superimposed (or weaved). The encryption filters ($[\![Send_e]\!]$, $[\![Receive_E]\!]$) are composed with the authentication filter ($[\![Auth_C]\!]$,$[\![Auth_S]\!]$), this is the *shared join point* whose composition we outlined in Sect. 5.4 and analyzed in Sect. 5.5.

The state machine specifications can be translated into Java code and defined in classes *authenticate* and *secureChannel* in package *dress4Less. SecurityProtocols*. This way security aspects can be first defined and formally analyzed, and later on weaved into a given program (or model) with the use of an aspect weaver such as COMPOSE* in the case of the code level. Fig. 5.5 illustrates the relation between model and code weaving as a complementary process. The base concern on the upper left part is either a component model, or a UML one, this is transformed by composing the security aspects on it. In the case of components by the composition operator defined in Sect. 5.1 and some pointcut designator as in Table 5.1 by defining a concern as in listing 5.1. Code generation can happen either before or after composing the aspects in the model. In the case we explored above, the UML model is translated into code and the state machines also, the weaving process is performed at the code level. The analysis of aspect interaction between the secure channel and authentication was performed at the model level and its suggested translation to code is proposed after composition (right hand side of Fig. 5.5).

Selecting the join point is based on the stereotypes «`high`» and «`encrypted`» In this case, when we have both of them on a communication

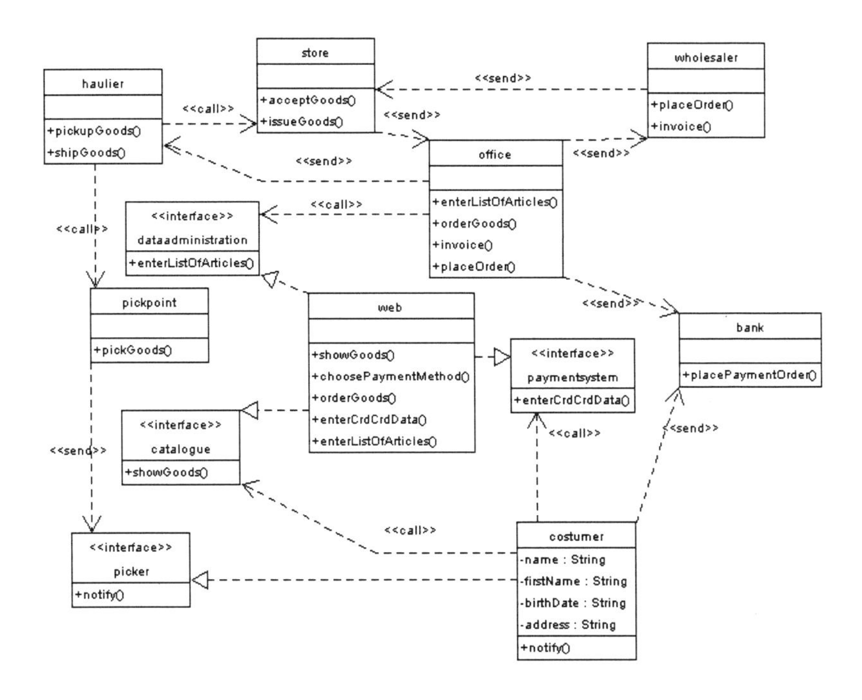

Figure 5.6: Class diagram web store

line between packages in the deployment diagram shown in Fig. 5.7, we select the payment related methods that are called upon. This is expressed in lines 3 through 4 of listing 5.1 (the complete concern definition can be found in Appendix E.1), specifically in the class diagram methods `placePaymentOrder` as well as `choosePaymentMethod` and `sendAuthenticated`. This one is the skeleton for the implementation of the authentication aspect with the encryption protocol. We may select other methods in classes such as `Costumer` if needed, based on the verification analysis or as a result of changing requirements. Weaving the secure channel is performed at the code level using COMPOSE* as pointcut delimiter (see listing 5.1). This way security aspects can be verified and used in case of software evolution, for instance.

5.8 Related Work

This chapter is a continuation of previous work where we considered analyzing crosscutting concerns on a formal theory [24].

Some related work was already mentioned in the introduction. We complete

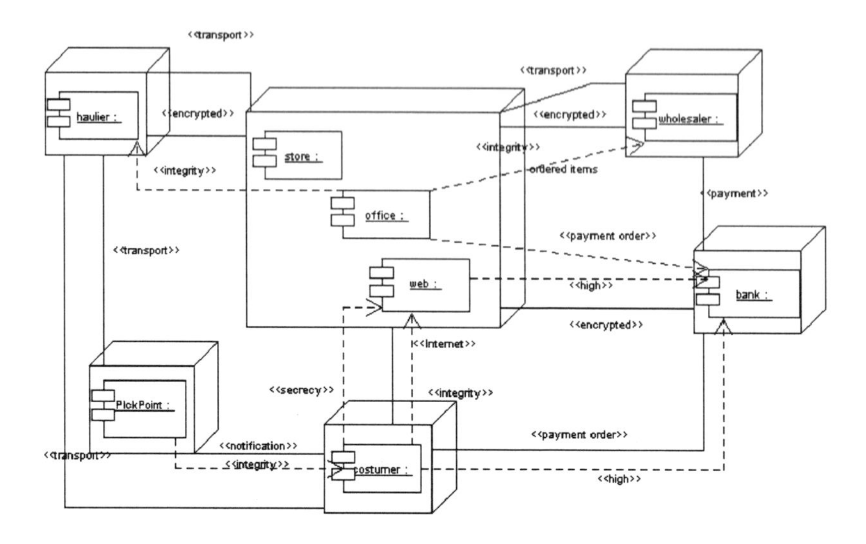

Figure 5.7: Deployment diagram with security specification stereotypes (web store)

the list here.

On a broader perspective the work of Compare et. al (see [14]) also focuses on an architecture level analysis for checking system properties. Our work in the tool framework has similar goals; we rely on certain extensions to UML (UMLsec) in order to establish the security properties the system has to fulfill. Additionally, our architectural model allows us to concentrate on aspect composition from a black box perspective. Haydar et. al (see [33]) also approach verification and validation from a black box perspective to formally model web applications and observe the external behavior of a web application yet our proposal is not specific for web applications. We also performed our analysis at the level of external behavior of components. Cottenier et. al [16] present excellent work related to ours in the sense that they also discuss aspects at the level of interface composition. The authors perform black box aspect composition represented as state machines. The authors tag state machines at certain decision points in SDL models to weave aspects. This allows them to follow the semantics of state machines and substitute a given execution pattern with a: before, after or around advice. Our framework is more directed toward static aspect weaving and emphasizes the use of the verification framework and the formal underpinnings for composition, i.e., weaving.

Nakajima and Tamai in [60] prove the use of a first order logic formal language (Alloy) for analysis of system design with respect to security requirements. They specify and analyze JAAS, the access control mechanism of JDK 1.2, with Alloy.

Listing 5.1: Selected lines from pointcut delimiter in COMPOSE*

```
...
inputfilters                                                                  2
 authenticate_filter : Dispatch = {True => [*. ↩
    placePaymentOrder]authentication.paytobankAuth,[*. ↩
    choosePaymentMethod]authentication.choosePaymentAuth}; ↩

 encrypt_filter : Dispatch = {True => [*.sendAuthenticated ↩               4
    ]encryption.encChannel}
}
superimposition                                                               6
{ selectors
payment = { *=dress4Less.Bank, *=dress4Less.Web, *= ↩                          8
   dress4Less.SecurityProtocols.authenticate};
filtermodules
 payment <- securePayment;                                                   10
}
```

We rely on a high order logic formal method (Focus) as supporting theory.

A survey on correctness and verification of aspect languages by the AOSD-Europe network of excellence can be found in [47]. Also the work of [57] represents an interesting approach to analysis of aspect interaction whereas its formal underpinnings are defined by means of graph transformation. A survey on a set of security patterns that can be found in the literature was published by [36].

5.9 Summary

In this chapter, we introduced a composition model that allows us to provide concerns with a (formal) syntax and semantics. In this, way we might be able to pose the problem of aspect composition with other aspects (the issue of aspect interaction) at a formal level. This work suggests that building a composition model on the concept of *I/O behavior* and the composition operators for *stream processing functions*, we may at least explore under which conditions aspects may be safely composed. This framework allowed us to consider aspect analysis at the semantic level which up to the present time has been mostly performed syntactically. Moreover, this work also proposed a generic secrecy enforcing channel (called secure channel through this work as explained in the introduction) that has been formally verified. Altogether, we introduced a formal model for composition filters and aspect interaction analysis, a generic secure channel aspect composed with an authentication protocol and an analysis of their interaction in view of preservation of security properties during aspect composition. We did not focus our attention on the issue of compositionality of security in the general case, as mentioned before, our aim was here to illustrate the framework at the hand of a more particular case.

However, one of the apparent limitations of this approach is the lack of analysis related to the pointcut designator or superimposition mechanism itself. In our proposal weaving is performed following the tables shown in this chapter at the selected channels, since we then obtain a new setting of components, we do not analyze the mechanism through which we select the base components, the weaving ordering indicated in the tables, since the resulting system represents the system or part we need to focus on. Its behavior, i.e., its semantics, are represented by the resulting composition and can be considered apart from the selection mechanism, since we consider it an intermediate step in the development process.

Chapter 6

Conclusions

You do not become a 'dissident' just because you decide one day to take up this most unusual career. You are thrown into it by your personal sense of responsibility, combined with a complex set of external circumstances. You are cast out of the existing structures and placed in a position of conflict with them. It begins as an attempt to do your work well, and ends with being branded an enemy of society. — Václav Havel, (Living in Truth, 1986)

The discussion in this thesis has given an in depth investigation on the topic of aspect orientation. Despite the popularity of AO and the number of tools, approaches and modeling techniques available to support one form or another of AO, regardless of some more or less precise definitions on *crosscuting* or *aspect*, there was no systematic study of the subject. Particularly, from a perspective independent of the programming level. There are certainly formalizations in the literature yet mostly, if not always, focused on the programming level. We pointed out that the subject itself cannot be defined simply by the programming constructs that embody what has been predominantly understood as aspect orientation, such as: *join points and advices*. These constructs are basically code transformation tools that together with some sort of composition mechanism, i.e., *weaving* allow enhancing a given base code.

Equally important, by exploring AO at the programming level we observed that the tools considered aspect oriented provide some sort of transformation mechanism that is applied on a given program using particular language constructs. This is also the case of aspect orientation at more abstract levels such as modeling. However, if we explore the subject in view of the question: how does aspect orientation make the world better?[1] we faced a difficult task answer-

[1]this question tortured the author all along the research process. Hence, the author is deeply in debt with his thesis director Prof. Dr. Dr. h.c. Manfred Broy, for he did not take a simple or unfounded answer to the questions posed. We even invited one of the main exponents in the field,

ing this, if we only defined the subject by the tools themselves.

Therefore, we approached the topic from a top down perspective, bearing also in mind the intuitions taken by surveying the programming level techniques. We identified two main issues as being at the core of aspect orientation: *crosscutting* and *weaving*. We define crosscutting at the hand of the transformation from (informal) requirements to subsequent stages of development, namely their translation to a design or specification. We concluded that no modularization entity may fully encapsulate the implementation of some requirements, and some features might be optional. Here appears the second "aspect" in AO, *weaving* as the composition mechanism that allows for defining a kind of abstraction layer over the base model or program and then inserting it at predefined points in the modules of the base model that should be enhanced. Observing crosscutting in view of a tracing or transformation from requirements to design or specifications, we provided a meaning to aspect orientation and the tools represented by it. We defined crosscutting as the cause and aspect as the solution to C^3, though it is not a silver bullet. We further identified two main groups of C^3: *Nonfunctional* and *Functional*. Since aspects are the implementation of C^3 we focused onto the second group: functional aspects.

Subsequently, a systematic overview was given to show that aspect weaving can be explained by a composition technique. This technique is based on communication refinement, it is namely a special form of behavior refinement.

Considered this way, (functional) aspect orientation is not quite a new concept. Though, it is performed in new ways and with constructs that were not available before aspect orientation was conceived in the mid 90's. The selection mechanisms of aspect orientation are also something rather novel. These were not present before and were also not considered in layered models as the reference model OSI/ISO we mentioned in chapter four.

Thereafter, we provided an overview of weaving from a mathematical perspective, proposed a formal composition model based on composition filters, and illustrated how it can be used to model aspects.

We also showed how the concepts of AOP can be explained at the software architecture level. Both mirror each other, despite that a top down perspective provides a foundation for a formal theory of aspects, say for instance, to explain aspect interaction. In this regard, we completed our work with an application of the theory by modeling the composition of two aspects and analyzing under which circumstances their composition preserves the specified properties of each one.

Prof. Dr Mehmet Aksit, to help us exploring the issue. Actually, Prof. Broy's way of pulling this thesis forward was in this authors opinion, most in concordance with the way Aristotle had done: "going to the essence of things"

Outlook The following lines of work come to our mind as we complete this stage of our research.

- First, we consider aspect interaction to be a continual open issue. Our framework can be regarded as a contribution in this direction, but also some related work we mentioned. Besides, we might also rely on lightweight modeling formalisms and model checking (or SAT solvers) to early detect aspect conflicts. Related to it is the issue of compositionality in a broader sense. The questions related to it have attracted the author's attention, but remain rather out of the scope of this work.

- Secondly, examining the role of the concepts that shape software modularization, say *components*, *classes*, *services*, etc. and their intrinsic limitations. It is actually a question related to *separation of concerns*.

- Thirdly, the definition of a formal model for join point selection or superimposition built as a transformation function at the model level.

Appendix A

Web Store. System Model

A.1 Problem Space: Use Cases

A.1.1 Use Case 1. ChooseProduct

Description: The customer navigates through the store catalogue and may select articles.
Related Actors: Customer, Office
Precondition: The web store is ready to take orders. The article catalogue is set, as well as prices and transportation costs.
Trigger: The customer selects the "search" or "browse" option in the catalogue.
Post condition: The customer has entered the articles he intends to buy in the shopping cart. The preliminary invoice is ready.
Standard Process:

1. The customer navigates the catalogue in search of clothes.
2. Once an article has been chosen, the customer selects size and color.
3. The customer confirms his selection.
4. The office registers the article in the shopping cart.
5. The customer continues shopping. Repeat steps 1 through 4 until the option "cash desk" or "abort" is selected.
6. The customer signals the end of his shopping by either the option "cash desk" or "abort".

   (a) By selecting the option "cash desk" the office prepares a preliminary invoice and sums the total payable (including transportation costs).
   (b) By aborting the shopping cart is cleaeed, the articles are de-registered and the main menu is shown with a message: shopping interrupted.

A.1.2 Use Case 2. DeliveryOptions

Description: The customer provides a delivery address or selects a delivery location (pickup point) from a list.
Related Actors: Customer, Office
Precondition: Article selection is closed. The shopping cart is not empty.
Trigger: Closing the shopping cart.
Post condition: The delivery address (i.e. pickup point) is set. The shopping cart is not empty. The client's information is registered (name(s), billing address, delivery address).
Standard Process:

1. The customer provides his name and billing address.
2. He selects one of two possible delivery options:
   (a) Delivery address. He provides it.
   (b) pickup point. He selects one from a list.
3. The office saves above mentioned information.

A.1.3 Use Case 3. PaymentProcess

Description: The customer gives his credit card number and the office verifies his account through the bank. The bank debits the amount due and transfers the money to the web store's account.
Related Actors: Customer, Office, Bank
Precondition: Shopping cart completed. Preliminary invoice is ready. Delivery Address or pickup point is set.
Trigger: Closing the delivery menu.
Post condition: Amount invoiced debited from the customer's account. This amount is transferred to the web store's account. Signal with "Okay" or "try again" sent to customer.
Standard Process:

1. The customer provides card number and related information in the web form.
2. The office sends these data to the bank, together with due amount.
3. The bank checks the card information.
   (a) If the information is correct, the amount due is debited from the customer's account. The web store becomes a confirmation and signals the customer "okay."
   (b) If the information is not correct, the bank sends a "try again" message. Up to three tries are allowed. Afterwards the process is aborted.

4. The office (web store) confirms the result of the process to the customer.

A.1.4 Use Case 4. ProcessOrder

Description: The shipper (transport company) prepares the delivery of goods to the client.
Related Actors: Shipper, Office
Precondition: The shipper is ready to attend shipping orders. The office has set the bill.
Trigger: The shipper receives a message from the office with the shipping information (article list and delivery address or pickup point).
Post condition: The shipper has the goods that are to be delivered. The delivery schedule is prepared.
Standard Process:

1. The office sends a message to the shipper concerning the recently confirmed deliveries.
2. The shipper reads the confirmed deliveries from the system (office) and prints them.
3. The shipper collects the required goods (from each delivery) from the company' warehouse.
4. The office updates the inventory, i.e., catalogue.
5. The office prepares a delivery schedule.
6. The shipper prints the delivery schedule and makes adjustments in case needed.

A.1.5 Use Case 5. DeliverGoods

Description: The shipper brings the goods (articles) to the desired address.
Related Actors: Shipper, Customer, Office
Precondition: Goods, Invoice, and Schedule are ready.
Trigger: The shipper accepts and prints the schedule.
Post condition: The customer receives goods and invoice.
Standard Process:

1. The shipper transports the goods according to the schedule.
2. On arrival by customer or pickup point the shipper registers the signature of the recipient.
3. The shipper registers in the system (office) the date and time of delivery.

A.1.6 Use Case 6. SetPrices

Description: The manager sets the profit margin and transportation costs. The system (office) sets the price of each article based on the base price from wholesaler and the profit margin.
Related Actors: Manager, Wholesaler
Precondition: Collection (Article, Price) is provided by the wholesaler.
Trigger: The wholesaler modifies the catalogue with a new collection.
Post condition: The web store's catalogue is complete.
Standard Process:

1. The wholesaler enters new articles (he provides size, price, quantity available, and color).
2. The office signals the manager that a new collection is available.
3. The manager selects articles for the web store's catalogue (in different sizes, colors, and quantities). He sets the catalogue of the web store by doing this.
4. The manager enters in the system (office) a global profit margin and delivery costs.
5. The office sets the price for the customer for each article (based on base price and profit margin).
6. Based on the manager's selection of goods, the office makes an advance order.
7. The manager checks this advance order and eventually reiterates steps 1 through 6 until the advance order is accepted.
8. The office sends the order to the wholesaler.

A.1.7 Use Case 7. ProvideGoods

Description: The wholesaler transports the goods to the web store.
Related Actors: Wholesaler, Office
Precondition: (Purchase) order is sent to the wholesaler.
Trigger: Incoming of the order.
Post condition: The articles are subtracted from the wholesaler's inventory and added to the web store's inventory (the catalogue that the customer navigates).
Standard Process:

1. The wholesaler delivers the goods to the web store.
2. The office updates the inventories. The goods are subtracted from the wholesaler's inventory and added to the web store's catalogue.

A.2 From Problem Space to Solution Space. Web Store Model

Use Case 2

We add attributes name and billing address to customer, make pickup point to a class and add the relationship choosing between it and the pickup point, which is managed as a composite of several pickup points (considered as predefined delivery addresses). See Fig. A.1.

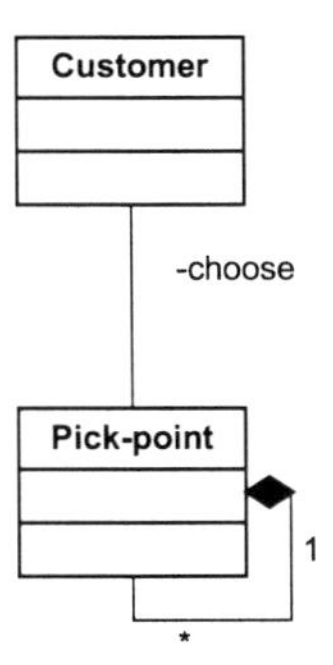

Figure A.1: Classes from UC2

Use Case 3

We identified the classes shown in Fig. A.2(a), however after reading the corresponding use case in search of attributes it became manifest that account should be treated as an abstract class with two concrete classes: . This was decided considering the difference between a checking account and a credit card account with respect to the attributes that each may contain. Therefore, class checking accounts have attributes regarding account numbers and balances, while credit card accounts have attributes regarding card numbers and credit limits. This exemplifies an iteration (as explained above) that enhances Fig. A.2(a) and results in Fig. A.2(b).

Use Case 4

Given that the office needs to update the inventories, we added an attribute quantity (qty) to a class article. Most other issues related to the elements in this use case have already been considered. The associated diagram is shown in Fig. A.3

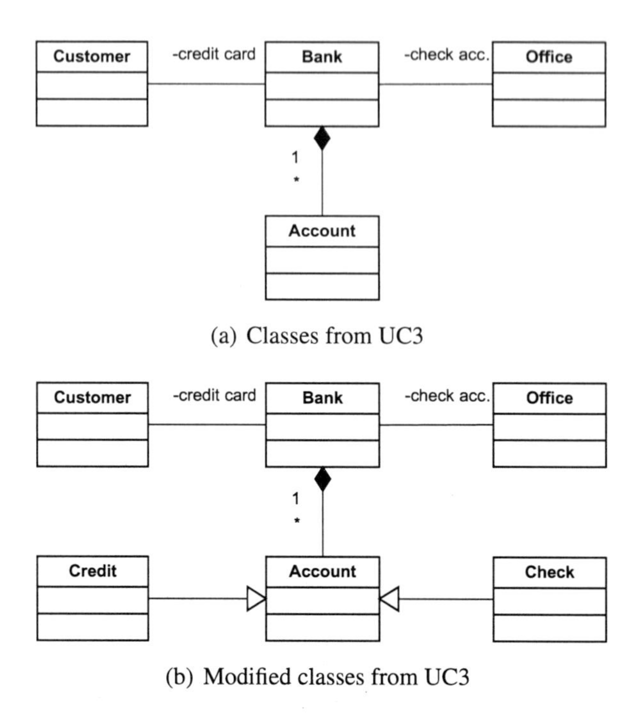

(a) Classes from UC3

(b) Modified classes from UC3

Figure A.2: Classes from UC3

Use Case 5

Class order is given the attributes delivery date, delivery time, and signature (customer's signature). Fig. A.4

Use Case 6

At the hand of this use case, we decided to make class catalogue into an abstract class with two generalizations: StoreCatalog and WhouseCatalog, in order to differentiate both inventories (web store's inventory and wholesaler's inventory). We recalled that in this use case the web store manager selects articles from the collection made available by the warehouse.

Class whousecat is related to the warehouse. Since the actor warehouse is mentioned in this and the next use case, this actor is made into the corresponding class. Class storecat is related to class office. These relationships are added in the global diagram (see Fig. A.7 in Appendix A.4). We will see later, that specifying the catalogue points to interesting design decisions when modeling in Alloy. In short, since we will define class attributes as sets (eventually as singleton sets), we may take advantage of some characteristics of the language, such as defining

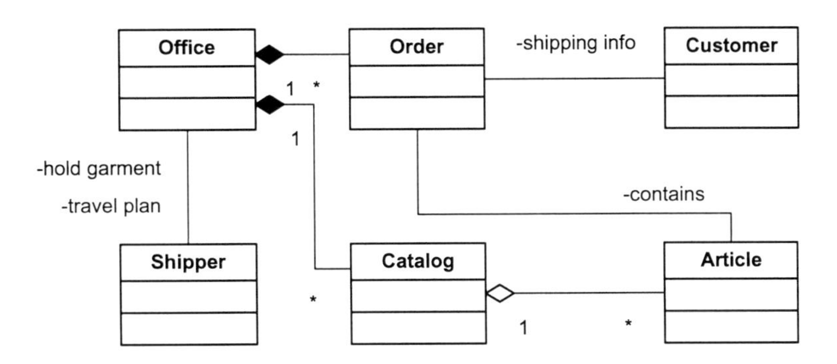

Figure A.3: Classes from UC4

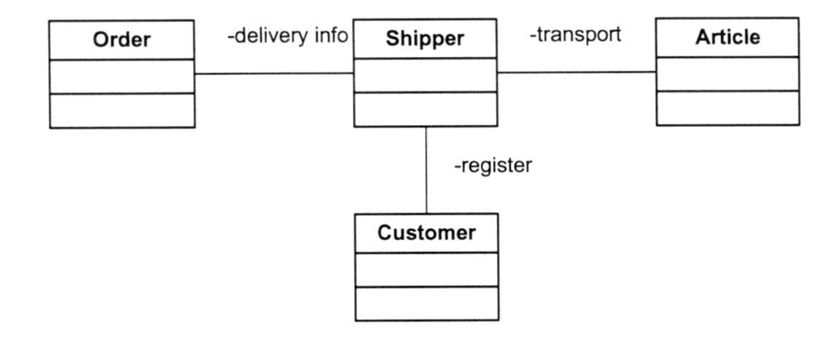

Figure A.4: Classes from UC5

disjoint sets helps to insure use case 7.

Back to use case 6, since the web store's manager provides a global profit margin and transportation costs. These attributes are allocated to class office. We also required adding class article with an attribute price. The last one will be used as base price if the article belongs to WhouseCat or as final price in case it belongs to StoreCat.

Use Case 7

This use case introduces the n-ary relation update/subtract/add article to/from the corresponding catalogue. See Fig. A.6.

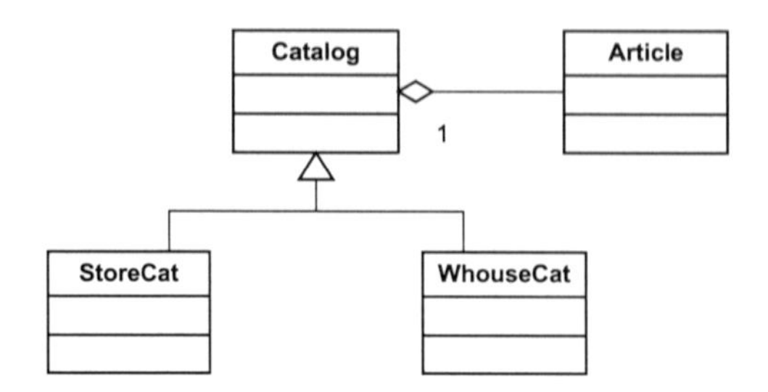

(a) Classes from UC6. Catalogue as abstract class with two generalizations

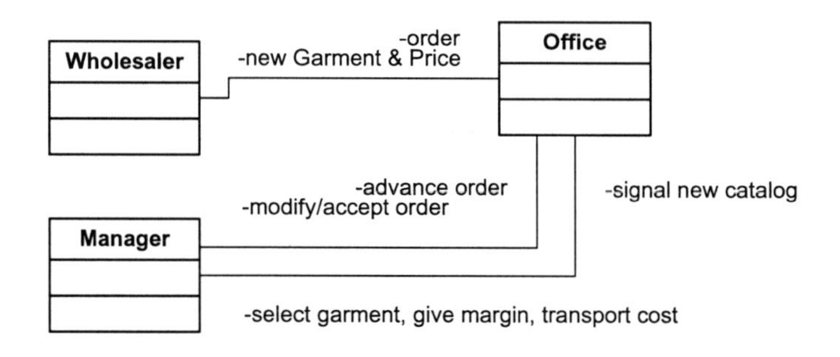

(b) Additional classes from UC6

Figure A.5: Classes from UC6

A.3 Solution Space: Specifications

Listing A.1: Structural specification for UC1

```
module thesis/WebStoreUC1

--
-- Use Case 1                                          4
--

sig Customer {
        name: Name,
        billingaddr: one Address,                      9
        deliveryaddr: one Address,
--      buy_sale: Office,
--      sortiment: lone ShoppingCart
}

sig Address {
} --see UC2, UC4, and 5
```

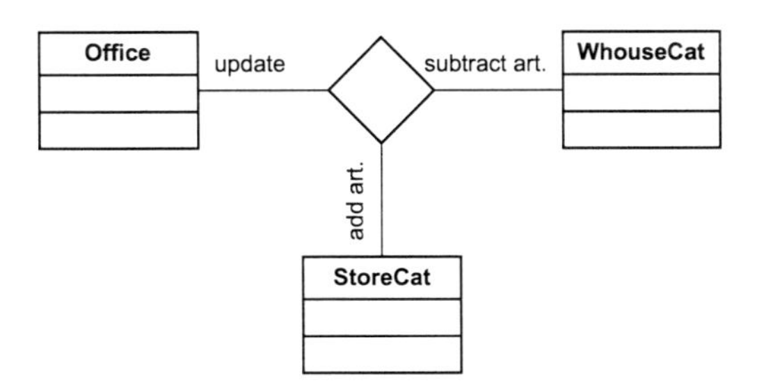

Figure A.6: Classes from UC7

```
sig Name {
} --see UC2                                                    19

one sig Office {
        cart: set ShoppingCart,
        pos: CashDesk,
--      cat: one Catalog                                       24
}

sig CashDesk {
}

sig Article {
        s: Size,
        c: Color,
        p:Price
}                                                              34

sig Size{
}

sig Color {                                                    39
}

sig Price {
}

/*
sig Catalog {
        contains: Office lone -> Article
}
*/                                                             49

sig StoreCat {
```

```
        contains: Office lone -> Article
}

sig ShoppingCart {
        contains: Customer lone -> Article
}

/*                                                          59
fun navigate (o:Office): Office->Article {
        o.cat.(contains)
}
*/

/*
pred naviCatalog(a: Article, o:Office) {
        a in o.cat
}
*/                                                          69

pred show () {
--      some r: Office -> Catalog | no iden & r
}

run show for 3 but 1 Catalog
run navigate --for 2 Office
```

Listing A.2: Structural specification for UC2

```
module thesis/WebStoreUC2
--
-- Use Case 2
--                                                          4
sig Customer {
        pickpoint: Customer -> one Pickpoint,
        deliveryAddr: Address
} -- fact forcing to have either one of Pickpoint or ↩
   deliveryAddres
--may be added here                                         9

sig Pickpoint {
        name: Name,
        address: one Address
}                                                           14

sig Address {
}

sig Name {                                                  19
}
```

```
pred show () {}
run show for 3
```

Listing A.3: Structural specification for UC3

```
module thesis/WebStoreUC3
--                                                          2
-- Use Case 3
--
sig Customer {
}

sig Office {
}

sig Bank {
        check_acc: Office -> Check,                         12
        credit_card: Customer -> Credit
}

sig Account {
        balance: one Balance,                               17
        pin: one PIN,
        accNumber: one Number
}

sig Check extends Account {                                 22
}

sig Credit extends Account {
        limit: Limit
}                                                           27

sig Number {
}

sig Limit {                                                 32
}

sig Balance {
}

sig PIN {
}

pred show () {}
run show for 3                                              42
```

Listing A.4: Structural specification for UC4 and UC5

```
module thesis/WebStoreUC4
--
-- Use Cases 4 and 5                                    3
--


sig Office {
        cat: lone Catalog,
        ord: set Order                                  8
}

sig Order {
        shipping_info: Customer  -> DeliveryInfo
}                                                       13

sig DeliveryInfo {
        address: Customer-> Address,
        date: Date,
        time: Time,                                     18
        sign: Signature
}

sig Customer {
}                                                       23

sig Shipper {
        holdgarment: Order -> Article,
        travelplan: Customer->Address
}                                                       28

sig Catalog {
        contains: Office lone -> Article
}

sig Article {
        qty: Quantity
}

sig Quantity {                                          38
}

sig Date {
}

sig Time {
}

sig Signature {
}                                                       48
```

```
sig Address {
}

pred show () {}                                                    53
run show for 3
```

Listing A.5: Structural specification for UC6

```
module thesis/WebStoreUC6                                           1
--
-- Use Case 6
--

sig Office {                                                        6
-- already defined in UC1
}

sig Catalog {
        contains: Office lone -> Article, --we may rename  ↩        11
            it to storecat
        whousecat: Wholesaler lone -> Article
}

sig Wholesaler {
}                                                                   16

sig Article {
}

pred show () {}                                                     21
run show for 3
```

A.4 Solution Space: Structural Model, Class Diagram

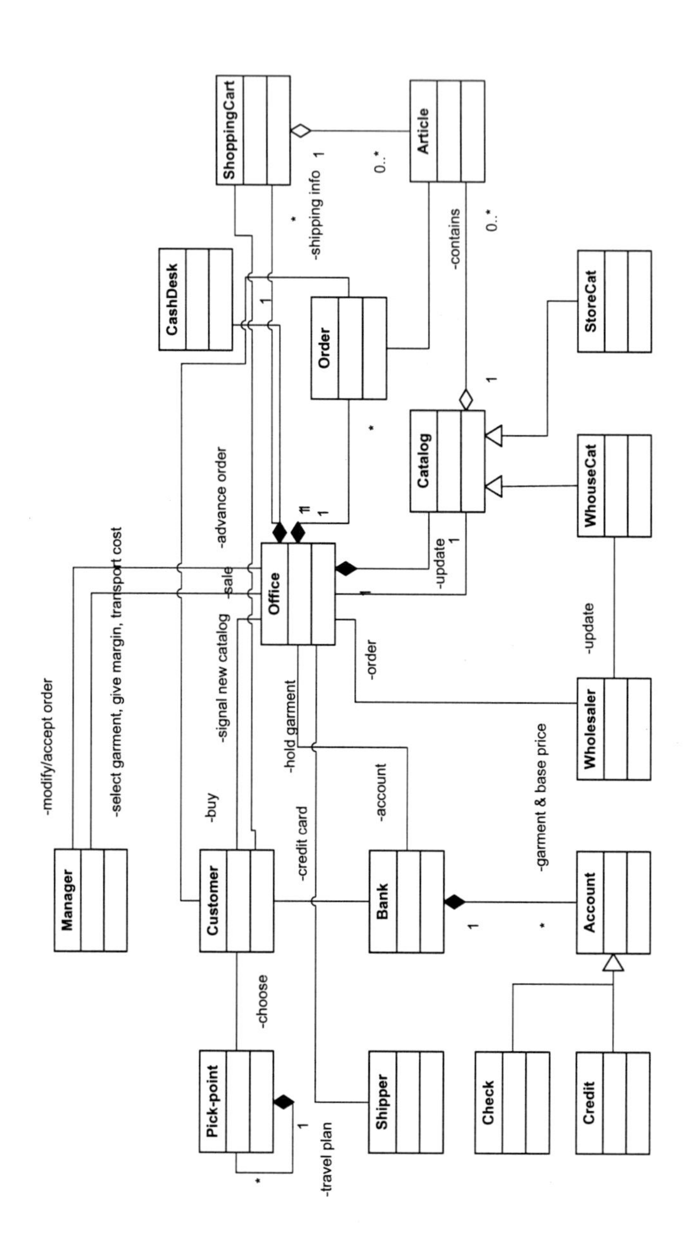

Figure A.7: Class Diagram of the Web Store. Alloy version

Appendix B

AspectJ Undo Example

Listing B.1: Class Circle. Cloning example

```
class Circle{
 private int rad;                                              2
 private int x=0;
 private int y=0;                                              4

 public Circle(int i) {                                        6
  rad = i;
 }                                                             8
 public Circle(int posx, int posy, int i) {
  x = posx;                                                    10
  y = posy;
  rad = i;                                                     12
 }
 public void moveCircle(int newposx, int newposy) {            14
 x = newposx;
 y = newposy;                                                  16
 }
 public int getradius() {                                      18
  return rad;
 }                                                             20
}
```

Listing B.2: Class Square. Cloning example

```
Class Square                                                   1
package clonning;

class Square {
 private int side;                                             5
 private int x=0;
 private int y=0;                                              7
```

```
  public Square(int i) {                                          9
    side = i;
  }                                                               11

  public Square(int posx, int posy, int i) {                      13
    x = posx;
    y = posy;                                                     15
    side = i;
  }                                                               17

  public void moveSquare(int newposx, int newposy) {              19
    x = newposx;
    y = newposy;                                                  21
  }

  public int getside() {
    return side;                                                  25
  }
}                                                                 27
```

Listing B.3: Introductions in ASPECTJ. Cloning

```
package clonning;                                                 1
import org.aspectj.lang.*;
import org.aspectj.lang.reflect.*;                                3

public aspect addCloneMethod {                                    5
  declare parents: Circle implements Cloneable;
  declare parents: Square implements Cloneable;                   7

  private int _indent = -1;                                       9

  public Object Circle.clone() {                                  11
    try {
      return super.clone();                                       13
    } catch (CloneNotSupportedException e)
    {                                                             15
      throw new InternalError(e.toString());
    }                                                             17
  }

  public Object Square.clone() {
    try {                                                         21
      return super.clone();
    } catch (CloneNotSupportedException e)                        23
    {
      throw new InternalError(e.toString());                      25
    }
  }                                                               27
```

```
pointcut tracePointsCircle(Circle c) : target(c) && call(* ↩    29
    moveCircle(..));

pointcut tracePointsCircle2(Circle c) : target(c) && set(* ↩    31
    Circle.*);

pointcut traceAllExecs() : !execution(* .new(..));               33

 void around (Circle c) : tracePointsCircle(c) {                  35
  .. actual implementation ...
 }                                                                 37

 void around (Circle c) : tracePointsCircle2(c) {                 39
  ... actual implementation ...
 }                                                                 41

 private void println(Object message) {                            43
  for (int i = 0, spaces = _indent * 2; i < spaces; ++i)
  {                                                                45
              System.out.print(" ");
  }                                                                47
  System.out.println(message);
 }                                                                 49
}
```

Listing B.4: Main program of the cloning example (ASPECTJ)

```
public class BasicClonning {
public static void main(String args[])throws ↩                   2
    CloneNotSupportedException {
  int size = 0;
  Square obj1 = new Square(27,33,4);                                4
  Square obj2 = (Square)obj1.clone();
  size = obj2.getside() << 1;                                       6
  System.out.println("Size of the Square = " + size);
  obj2.moveSquare(22,28);                                           8
  Circle obj3 = new Circle(25,30,4);
  System.out.println("Object " + obj3.toString()+                  10
  " posX: " + obj3.circlePosx() + " " + "posY: " + obj3. ↩
     circlePosy());
  Circle obj4 = (Circle)obj3.clone();                               12
  // just to try the introduction of method clone
  System.out.println("Object " + obj4.toString()+ " posX: " ↩     14
      + obj4.circlePosx() + " "
  +"posY: " + obj4.circlePosy());
  obj4.moveCircle(25,23);                                           16
  System.out.println("Object " + obj4.toString()+ " posX: " ↩
      + obj4.circlePosx() + " " +
```

```
    "posY: " + obj4.circlePosy());                          18
  }
}                                                           20
```

Appendix C

AspectJ Web Payment Example

C.1 Base Classes

Listing C.1: Bank payment example. Class bank

```
package dress4Less;

public class Bank extends Thread
{                                                                    4

 private String name;                                                6
 public Bank(String bname)
 {                                                                   8
  this.name = bname;
 }                                                                   10

...                                                                  12

 public int placePaymentOrder(String who, Costumer cx)               14
 {
  System.out.println(this.name+ " processing order  "+ who+ ↩       16
      "<->"+cx.getCostumerName());
  return 0;
 } // end placePaymentOrder                                          18
} // end bank
```

Listing C.2: Bank payment example. Class web

```
package dress4Less;                                                  1

public class Web //implements Catalogue, Paymentsystem,  ↩           3
    Dataadministration
{
```

```
public void choosePaymentMethod(String sx)                          7
{
 System.out.println("Web >Payment Method: credit card, ↩            9
    client:"+sx);
} // end choosePaymentMethod

public void orderGoods(Costumer cx, Bank bx)
{                                                                   13
 System.out.println(cx.getCostumerName() + " ordering ↩
    goods");
 bx.placePaymentOrder("web",cx);                                    15
} // end orderGoods

public int enterCrdCrdData(Costumer cx, Bank bx)
{                                                                   19
 System.out.println("Web >"+cx.getCostumerName() +" "+" ↩
    Credit card No. " +cx.getCrdNumber());
 return bx.placePaymentOrder("web",cx);                             21
} // end enterCrdCrdData
} // end web                                                        23
```

Listing C.3: Bank payment example. Class office

```
package dress4Less;                                                 1

public class Office {                                               3

 private int office_number;                                         5
 private String address;

 public Office(int i, String addr)
 {                                                                  9
  this.office_number = i;
  this.address = addr;                                              11
 }
...                                                                 13

 public void placeOrder(Bank bank, Costumer cx)                     15
 {
  bank.placePaymentOrder(this.address, cx);                         17
 } // end placeOrder

} // end office
```

Listing C.4: Bank payment example. Class customer

```
package dress4Less;
```

```
public class Costumer extends Thread
{                                                                    4
 private String name;
 private String firstName;                                           6
 private int crdNumber;
 private String birthDate;                                           8
 private String address;

public Costumer(String name1, String firstname1, int CCard, ↩
    String birthdate1, String address1)
 {                                                                   12
  this.name = name1;
  this.firstName = firstname1;                                       14
  this.birthDate = birthdate1;
  this.address = address1;                                           16
  this.crdNumber = CCard;
 }                                                                   18

...                                                                  20

public void buyGoods(Web w, Bank b)                                  22
{
 String t;                                                           24
 t=this.getCostumerName();
 System.out.println(t+"> "+"buying at dress4Less");                  26
 w.choosePaymentMethod(this.getCostumerName());
 if (w.enterCrdCrdData(this, b)==0){                                 28
  System.out.println(this.name+ " "+this.firstName+ " ↩
     Confirmation received");
 }                                                                   30
 else
 {                                                                   32
  System.out.println(this.name+ " "+this.firstName+" ↩
     failure somewhere");
 }                                                                   34
}


public void paytobank(Bank b1)                                       38
{
 System.out.println(this.name + " " + this.firstName+ " "+ ↩         40
    "paying via :" +b1.getBankname());
}
} // end costumer                                                    42
```

C.2 Aspects

Listing C.5: Bank payment example. Before advice for payment()

```
package dress4Less;

import org.aspectj.lang.JoinPoint;

public aspect FakeAuthenticate {
 public pointcut tracePayment(Bank bank): call(int Bank. ↩       6
    placePaymentOrder(..)) && target(bank);

 before (Bank b) : tracePayment(b)                                8
 {
  System.out.println("before " + thisJoinPoint + " ↩              10
     Authenticate here");
 }
}                                                                 12
```

Please note that in "before" advices the control flow returns to the intercepted activating event so that it may actually be performed. Therefore, in this particular example the authentication does not need to send a confirmation to the activating event (the call to method payment), because the method payment is not substituted (though a solution could also be, to use an *around* aspect and then decide from the aspect itself when the activating event is called). Though, an adequate "error" handling mechanism should be considered in case the authentication fails, our simple example clearly does not take this situation into consideration.

Listing C.6: Bank payment example. After advice for payment()

```
package dress4Less;

public aspect NotifyWarehouse {
 public pointcut tracePayment(Bank bank): call(int Bank. ↩       4
    placePaymentOrder(..)) && target(bank);

 after (Bank b) : tracePayment(b)                                 6
 {
  System.out.println("after " + thisJoinPoint + " Notify ↩        8
     warehouse");
 }
}                                                                 10
```

Listing C.7: Bank payment example. Around advice for payment()

```
package dress4Less;

public aspect otherPayment {
```

```
public pointcut tracePayment(Bank bank): execution(int *. ↩          4
    placePaymentOrder(..)) && this(bank);

int around (Bank b) : tracePayment(b){                                 6
System.out.println("otherPayment> Mail order");
return 1;                                                              8
}
}                                                                      10
```

Appendix D

E-Learning Support System: Requirements

D.1 Problem Space: Stakeholders, Interviews, Requirements

D.1.1 Problem Description

During the spring of 2002 in the framework of a Data Warehousing and Mining training program in Hyderabad, India, we designed an E-Learning Support System (ELSS) for a Mexican university. The university had at that time had no distance learning support system.

The motivation was the following. Given the location of the university at the heart of Mexico City, many students spend a considerable amount of time traveling to and from the university. Some students come to the university from different parts of the city and its metropolitan area, some even from neighboring cities.

During the training program in India the requirements' analysis was not documented, because the focus then was to develop a prototype within a short span of two weeks. The author made a prototype using Visual Basic for the front end and SQL Server for the back end. A technical report can be found in [23].

We refer to the ELSS as case study for requirements. For this purpose, an informal description of its requirements follows.

First, we outline the stakeholders we considered at that time. Secondly, we devise a kind of interview simulating a real set of interviews from which we may then elaborate a list of requirements.

D.1.2 Stakeholders

Academic Personnel These are the teachers of the language center as well as the director of the center. The academic personnel give lectures and monitor the students' progress regularly, based on the university's calendar and the course learning objectives. The director of the language center coordinates the learning process, resolves conflicts that may arise from group allocation, teacher student interaction; supervises the evaluation process, reports advances to the academic direction of the university and sometimes also gives lessons.

Students These are the individuals registered at the university in a given course or educational program. We concentrate on the students attending English language courses. Their advance is regularly evaluated in terms of mastering a given language skill. Students of the language center are registered at different academic faculties and share heterogeneous professional and cultural interests.

Administrative Personnel These are the individuals in charge of registering students, printing out certificates and supporting the activities of the members of the university.

D.1.3 Shaping the Requirements. Interviews

We chose to provide simulated interviews as a way of introducing the requirements in textual form. The following interviews have been designed by an expert in this domain, namely the author itself. The names are fictitious.

Miss Sylvia (Teacher) "I would like to grade exams on a periodical basis and obtain the results in a clear and understandable way. I would like to enter the correct answers of a particular evaluation in the system and obtain the number of correct answers per student and group. An interesting goal would be to automatically obtain comparisons among groups and identify questions in which the students have had the most difficulties in answering would be helpful. This would certainly help us improve the learning process."

Miss Torvax (Director of the Language Center) "There is a lot of work to do here. Sometimes, especially at the end of the semester, we need to correct exams, give the final notes to the registrar and submit a report to the Dean. During the semester, we also have to keep track of the evolution of the groups. At the beginning of the semester, I am in charge of assigning teachers and groups, according to availability. This is no easy task. Perhaps the new system should leave such a complex problem for a later edition. Now, let us focus on the evaluation aspect of the problem."

Miss Torvax also told us that the typical controls and final exams very rarely exceed a number of forty multiple choice questions. She allowed us to visit an exam session which provided us with valuable insights of the university's procedures.

David (Student) "Well,...I'd like to review additional material at home and check my advancement at my own pace. As I live outside the city, taking exams at home would be nice. I'd also like to read the lecture notes on the web."

Sara (Student) "I would feel so happy if I could present the examinations in advance, because sometimes I am ready to take an exam weeks before the testing dates."

D.1.4 Design Requirements

The administrative personnel made a particular recommendation for this software. They suggested that software produced for the ELSS should be composed of the following three parts:

1. A relational database (DB) implemented on *MySQL* as *back end.* This part will be further referred to as DB.

2. A front end Java application for evaluations and automatically reporting students' results based on the correct answers stored in the DB. This is the Exams module. The front end should also contain an administration module.

3. A web site with course supporting material for students should also be provided.

D.1.5 List of Requirements for the ELSS

1. Functional Requirements

   (a) Allow students to study course related support material on the internet.

   (b) Provide distance lessons.

   (c) Allow students to take exams on selected terminals in the campus.

   (d) Allow students to take exams at their own pace and when they feel confident enough to. The former has to be coordinated with the university's semester calendar.

2. Additional Requirements

   (a) Access to the system shall be granted only to authorized users.
   (b) Identity of users must be certified.
   (c) Only Academic Personnel is allowed to create, modify, or delete tests or evaluations, in the courses they are in charge of.
   (d) Only Administrative Personnel may register a Student for a course and assign courses to Academic Personnel.
   (e) Student information should be kept confidential, specially grades.

3. Extra Functional Requirements

   (a) Availability: the system should be available 24 hours a day, 7 days a week, though the system might stay off line until any fault is fixed.
   (b) Performance: the system shall be able to attend at least 100 users simultaneously

D.2 Solution Space: ER Diagram, Artifacts, Class Diagram

D.2.1 Artifacts: Entity-Relationship SQL Diagram and Screens

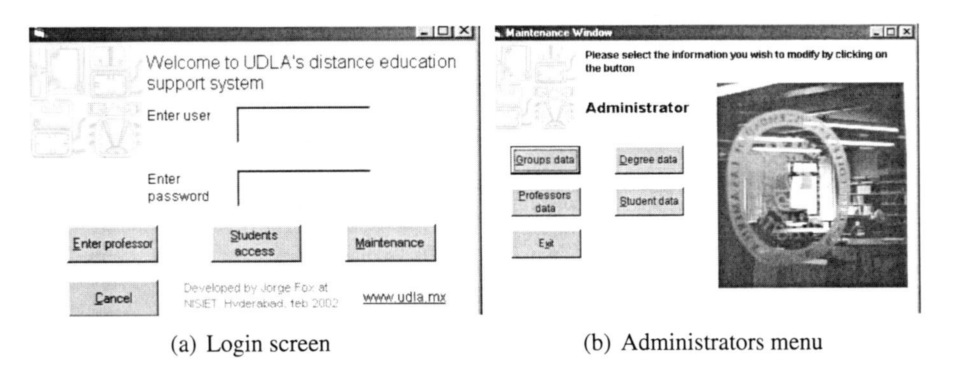

(a) Login screen (b) Administrators menu

Figure D.1: ELSS sample screens

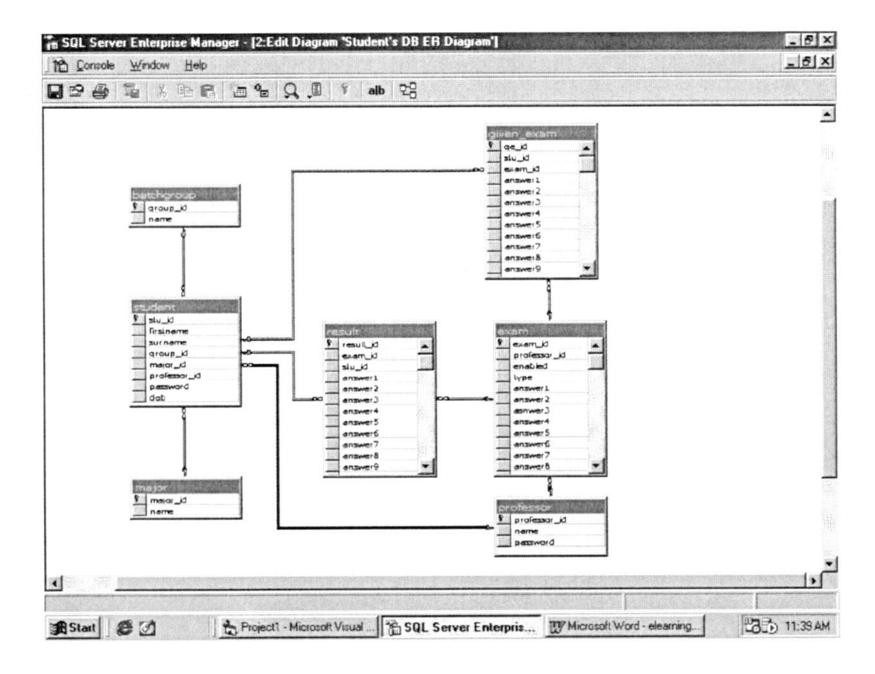

Figure D.2: ER diagram. ELSS

D.2.2 ELSS Class Diagram

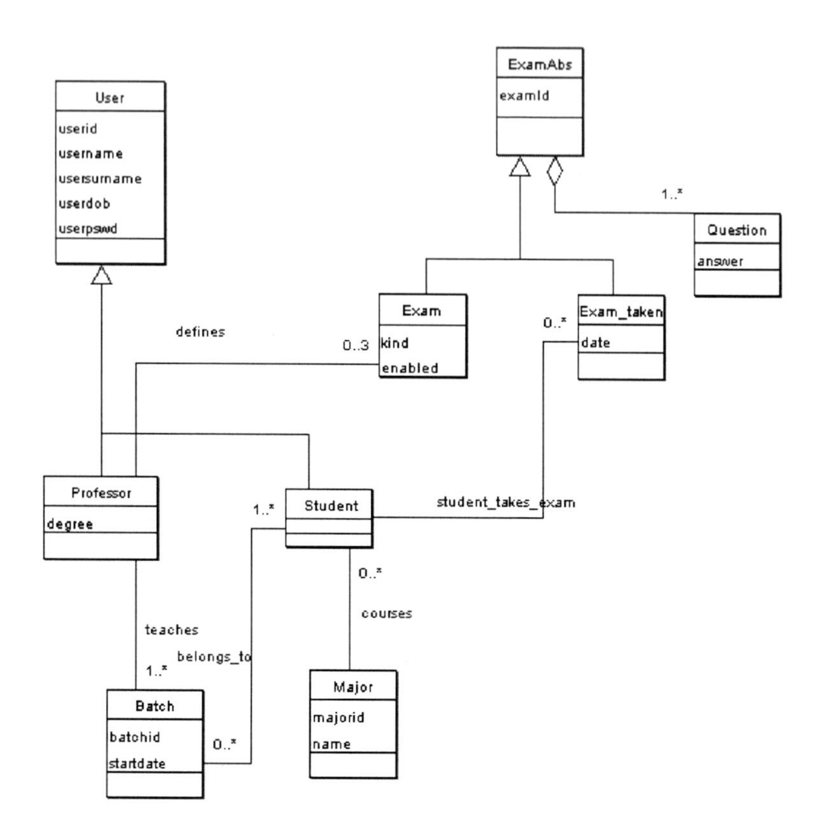

Figure D.3: Class diagram ELSS

Appendix E

Web Store Design

E.1 COMPOSE* Filters. Code Skeletons

Listing E.1: Code skeleton for the authentication filter

```
package dress4Less.SecurityProtocols;

import dress4Less.*;

public class authenticate
{                                                          6
        public authenticate()
        {                                                  8

        }                                                  10

public void placePaymentOrder(String sname)               12
        {
                try                                        14
                {
                        System.out.println("here:  ↩      16
                            Authenticate.. payment order  ↩
                            from :" +sname);
                }
                catch (Exception e)                        18
                {
                        System.out.println("Caught in Main: ↩   20
                                "+e);
                        e.printStackTrace();
                }                                          22
        }

        public void paytobankAuth(String sx)
        {                                                  26
```

```
        System.out.println("here: Authenticate.. ↩
           payment order from :" +sx);
        sendAuthenticated(sx);                              28
    }

    public void choosePaymentAuth(String sx)
    {                                                       32
        System.out.println("here: Authenticate.. ↩
           choose payment method, client:"+sx);
        sendAuthenticated(sx);                              34
    }

    public void sendAuthenticated(String data)
    {                                                       38
        System.out.println("here: comm channel ↩
           Authenticated client >"+data+ " <to ↩
           server");
    }                                                       40

}                                                           42
```

Listing E.2: Code skeleton for the secure channel filter

```
package dress4Less.SecurityProtocols;

import dress4Less.*;

/**
 * Summary description for secureChannel.                   6
 */
public class secureChannel                                  8
{
    public secureChannel()                                  10
    {

    }

    public void encryptSend(String data)
    {                                                       16
        System.out.println("channel sending:"+data) ↩
           ;
    }                                                       18

    public void encryptReceive(String data)                 20
    {
        System.out.println("channel receiving:"+ ↩          22
           data);

    }                                                       24
```

```
    public void encChannel(String data)
    {                                                         26
            this.encryptSend("channel s"+data);
            this.encryptReceive("channel r"+data);            28
    }

}
```

Bibliography

[1] Merriam webster online dictionary. http://www.m-w.com.

[2] Martín Abadi and Leslie Lamport. Composing specifications. *ACM Trans. Program. Lang. Syst.*, 15(1):73–132, 1993.

[3] Mehmet Akşit, Ken Wakita, Jan Bosch, Lodewijk Bergmans, and Akinori Yonezawa. Abstracting object-interactions using composition-filters. In R. Guerraoui, O. Nierstrasz, and M. Riveill, editors, *Object-Based Distributed Processing*, pages 152–184. svlncs, 1993.

[4] Mehmet Aksit. Composition and separation of concerns in the object-oriented model. *ACM Computing Surveys*, 28A(4), 1996.

[5] Colin Atkinson and Thomas Kühne. Aspect-oriented development with stratified frameworks. *IEEE Software*, 20(1):81–89, January/February 2003.

[6] Lodewijk Bergmans. Towards detection of semantic conflicts between crosscutting concerns. In Jan Hannemann, Ruzanna Chitchyan, and Awais Rashid, editors, *Workshop on Analysis of Aspect-Oriented Software*, ECOOP 2003, 2003.

[7] Lodewijk Bergmans, Mehmet Akşit, and Bedir Tekinerdogan. Aspect composition using Composition Filters. In Mehmet Akşit, editor, *Software Architectures and Component Technology*, pages 357–382. Kluwer Academic Publishers, 2001.

[8] Manfred Broy. Compositional refinement of interactive systems modelled by relations. In Willem-Paul de Roever, Hans Langmaack, and Amir Pnueli, editors, *Compositionality: The Significant Difference*, LNCS 1536, pages 130–149. Springer, 1998.

[9] Manfred Broy. *From States to Histories: Relating States and History Views onto Systems*, volume 180 of *Springer NATO ASI Series*, pages 149 – 186. IOS Press, 2001.

[10] Manfred Broy, Anton Deimel, Juergen Henn, Kai Koskimies, Frantisek Plasil, Gustav Pomberger, Wolfgang Pree, Michael Stal, and Clemens Szyperski. What characterizes a (software) component? *Software – Concepts & Tools*, 19(1):49–56, 1998.

[11] Manfred Broy and Ketil Stoelen. *Specification and Development of Interactive Systems: Focus on Streams, Interfaces, and Refinement*. Springer, 2001.

[12] Siobhán Clarke. *Composition of Object-Oriented Software Design Models*. PhD thesis, Dublin City University, 2001.

[13] Curt Clifton. *A design discipline and language features for modular reasoning in aspect-oriented programs*. PhD thesis, Iowa State University, 2005. ISU TR 05-15.

[14] Daniele Compare, Paola Inverardi, Patrizio Pelliccione, and Alessandra Sebastiani. Integrating model-checking architectural analysis and validation in a real software life-cycle. In Keijiro Araki, Stefania Gnesi, and Dino Mandrioli, editors, *FME*, volume 2805 of *Lecture Notes in Computer Science*, pages 114–132. Springer, 2003.

[15] Composestar project. http://composestar.sf.net.

[16] Thomas Cottenier, Aswin van den Berg, and Tzilla Elrad. Joinpoint inference from behavioral specification to implementation. In *Proceedings of the 21st European Conference on Object-Oriented Programming*, to appear. 2007.

[17] Krzysztof Czarnecki and Ulrich W. Eisenecker. *Generative Programming - Methods, Tools, and Applications*. Addison-Wesley, May 2000.

[18] P. E. A. Durr, L. M. J. Bergmans, and M. Akşit. Reasoning about behavioral conflicts between aspects. Technical Report TR-CTIT-07-15, Centre for Telematics and Information Technology, University of Twente, Enschede, February 2007.

[19] Tzilla Elrad, Mehmet Akşit, Gregor Kiczales, Karl Lieberherr, and Harold Ossher. Discussing aspects of AOP. *Commun. ACM*, 44(10):33–38, 2001.

[20] Tzilla Elrad, Robert E. Filman, and Atef Bader. Aspect-oriented programming: Introduction. *Communications of the ACM*, 44(10):29–32, 2001.

[21] R. Filman, T. Elrad, S. Clarke, and M. Akşit, editors. *Aspect-Oriented Software Development*. Addison-Wesley, 2004.

[22] Robert E. Filman and Daniel P. Friedman. Aspect-oriented programming is quantification and obliviousness. In R. Filman, T. Elrad, S. Clarke, and M. Akşit, editors, *Aspect-Oriented Software Development*, pages 21–35. Addison-Wesley, 2004.

[23] Jorge Fox. E-learning support system. Technical report, National Institute of Small Industry Extension Training (nisiet), Hyderabad, India, March 2002.

[24] Jorge Fox. A formal foundation for aspect-oriented software development. *Research on Computing Science, CIC-IPN, ISSN: 1665-9899*, 14:241–251, 2005.

[25] Jorge Fox. A taxonomy of aspects in terms of crosscutting concerns. In Ed Brinksma, David Harel, Angelika Mader, Perdita Stevens, and Roel Wieringa, editors, *Methods for Modelling Software Systems (MMOSS)*, number 06351 in Dagstuhl Seminar Proceedings. Internationales Begegnungs- und Forschungszentrum für Informatik, 2007.

[26] Jorge Fox and Jan Jürjens. Introducing security aspects with model transformation. In *Model Based Development (MBD) Workshop in Proc. 12th Annual IEEE International Conference on the Engineering of Computer-Based Systems (ECBS 2005), Greenbelt, Washington, 4-5 April*, pages 543 – 549. IEEE Computer Society, 2005.

[27] Jorge Fox and Jan Jürjens. A framework for analyzing composition of security aspects. In Ed Brinksma, David Harel, Angelika Mader, Perdita Stevens, and Roel Wieringa, editors, *Methods for Modelling Software Systems (MMOSS)*, number 06351 in Dagstuhl Seminar Proceedings. Internationales Begegnungs- und Forschungszentrum für Informatik (IBFI), 2007.

[28] Joseph Gradecki and Nicolas Lesiecki. *Mastering AspectJ*. Wiley Ed., 2003.

[29] Johannes Grünbauer, Helia Hollmann, Jan Jürjens, and Guido Wimmel. Modelling and verification of layered security protocols: A bank application. In *SAFECOMP 2003, 23-26 September 2003 Edinburgh, GB*. Springer-Verlag, 2003.

[30] Stefan Hanenberg, Dominik Stein, and Rainer Unland. Eine taxonomie für aspektorientierte systeme. In Peter Liggesmeyer, Klaus Pohl, and Michael Goedicke, editors, *Software Engineering*, volume 64 of *LNI*, pages 167–178. GI, 2005.

[31] David Harel and Amir Pnueli. On the development of reactive systems. In K.R. Apt, editor, *Logics and Models of Concurrent Systems*, volume F-13 of

Springer NATO ASI Series, pages 477–498. Springer-Verlag, New York, NY, USA, 1985.

[32] Wilke Havinga, Istvan Nagy, Lodewijk Bergmans, and Mehmet Akşit. A graph-based approach to modeling and detecting composition conflicts related to introductions. In *AOSD '07: Proceedings of the 6th international conference on Aspect-oriented software development, Vancouver, Canada*, pages 85–95, New York, NY, USA, 2007. ACM Press.

[33] May Haydar, Alexandre Petrenko, and Houari A. Sahraoui. Formal verification of web applications modeled by communicating automata. In David de Frutos-Escrig and Manuel Núñez, editors, *FORTE*, volume 3235 of *Lecture Notes in Computer Science*, pages 115–132. Springer, 2004.

[34] Dominikus Herzberg. *Modeling Telecommunication Systems: From Standards to System Architectures*. PhD thesis, Aachen University of Technology, Department of Computer Science III, 2003.

[35] Dominikus Herzberg and Manfred Broy. Modeling layered distributed communication systems. *Formal Aspects of Computing*, 17(1):1–18, 2005.

[36] Thomas Heyman, Koen Yskout, Riccardo Scandariato, and Wouter Joosen. An analysis of the security patterns landscape. In *SESS '07: Proceedings of the Third International Workshop on Software Engineering for Secure Systems*, page 3, Washington, DC, USA, 2007. IEEE Computer Society.

[37] Institute of Electrical and Electronics Engineers. IEEE standard computer dictionary: A compilation of IEEE standard computer glossaries, 1990.

[38] Anna Ioshpe. Anwendung modellbasierter sicherheitsanalyse. Systementwicklungsprojekt, Technische Universität München, Januar 2004.

[39] Daniel Jackson. Alloy: A logical modelling language. In Didier Bert, Jonathan P. Bowen, Steve King, and Marina A. Waldén, editors, *ZB*, volume 2651 of *Lecture Notes in Computer Science*, page 1. Springer, 2003.

[40] Daniel Jackson. *Software Abstractions. Logic, Language, and Analysis*. MIT Press, 2006.

[41] Michael Jackson. Problem frames and software engineering. *Information and Software Technology*, 47(14):903–912, November 2005.

[42] Ivar Jacobson. Use cases and aspects -working seamlessly together. *Journal of Object Technology*, pages 7–28, 2003.

[43] Jan Jürjens. *Secure Systems Development with UML*. Springer-Verlag, 2005.

[44] Jan Jürjens. Sound methods and effective tools for model-based security engineering with UML. In *27th International Conference on Software Engineering (ICSE 2005)*, 2005.

[45] Jan Jürjens and Jorge Fox. Tools for model-based security engineering. In *ICSE '06: Proceeding of the 28th international conference on Software engineering*, pages 819–822, New York, NY, USA, 2006. ACM Press.

[46] Jan Jürjens and Pasha Shabalin. Automated verification of UMLsec models for security requirements. In J.-M. Jézéquel, H. Hußmann, and S. Cook, editors, *UML 2004 – The Unified Modeling Language*, volume 2460, pages 412–425, 2004.

[47] Shmuel Katz. A survey of verification and static analysis for aspects. AOSD-Europe-Technion-1, AOSD-Europe, July 2005.

[48] Shmuel Katz. Aspect categories and classes of temporal properties. In Awais Rashid and Mehmet Akşit, editors, *T. Aspect-Oriented Software Development I*, volume 3880 of *Lecture Notes in Computer Science*, pages 106–134. Springer, 2006.

[49] Gregor Kiczales and Erik Hilsdale. Aspect-oriented programming. In *ESEC/FSE-9: Proceedings of the 8th European software engineering conference held jointly with 9th ACM SIGSOFT international symposium on Foundations of software engineering*, page 313, New York, NY, USA, 2001. ACM Press.

[50] Gregor Kiczales, Erik Hilsdale, Jim Hugunin, Mik Kersten, Jeffrey Palm, and William Griswold. Getting started with AspectJ. *Communications of the ACM*, 44(10):59–65, 2001.

[51] Gregor Kiczales, John Lamping, Anurag Menhdhekar, Chris Maeda, Cristina Lopes, Jean-Marc Loingtier, and John Irwin. Aspect-oriented programming. In Mehmet Akşit and Satoshi Matsuoka, editors, *Proceedings European Conference on Object-Oriented Programming*, volume 1241, pages 220–242. Springer-Verlag, Berlin, Heidelberg, and New York, 1997.

[52] Gregor Kiczales and Mira Mezini. Aspect-oriented programming and modular reasoning. In *ICSE '05: Proceedings of the 27th international conference on Software engineering*, pages 49–58, New York, NY, USA, 2005. ACM Press.

[53] Ivan Kiselev. *Aspect-Oriented Programming with AspectJ*. Sams, Indianapolis, IN, USA, 2002.

[54] Karl Lieberherr. Adaptive software 93/94 research report. Technical report, College of Computer Science, 1994. www.ccs.neu.edu/research/demeter/aop/history/93-94-report.

[55] Karl Lieberherr, Doug Orleans, and Johan Ovlinger. Aspect-Oriented Programming with Adaptive Methods. Technical Report NU-CCS-2001-01, College of Computer Science, Northeastern University, Boston, MA, February 2001.

[56] Marius Marin, Leon Moonen, and Arie van Deursen. A common framework for aspect mining based on crosscutting concern sorts. Technical Report TUD-SERG-2006-009, Delft University of Technology, 2006.

[57] Katharina Mehner, Mattia Monga, and Gabriele Taentzer. Interaction analysis in aspect-oriented models. In *RE*, pages 66–75. IEEE Computer Society, 2006.

[58] Haralambos Mouratidis, Jan Jürjens, and Jorge Fox. Towards a comprehensive framework for secure systems development. In Eric Dubois and Klaus Pohl, editors, *Proceedings of the 18th International Conference, CAiSE 2006*, volume 4001, pages 48–62. Springer Berlin / Heidelberg, June 5-9 2006.

[59] Istvan Nagy, Lodewijk Bergmans, and Mehmet Akşit. Composing aspects at shared join points. In Robert Hirschfeld, Ryszard Kowalczyk, Andreas Polze and Mathias Weske, editor, *Proceedings of International Conference NetObjectDays, NODe2005*, volume P-69 of *Lecture Notes in Informatics*, Erfurt, Germany, Sep 2005.

[60] Shin Nakajima and Tetsuo Tamai. Formal specification and analysis of JAAS framework. In *SESS '06: Proceedings of the 2006 international workshop on Software engineering for secure systems*, pages 59–64, New York, NY, USA, 2006. ACM Press.

[61] Netbeans project. Open source. Available from http://mdr.netbeans.org, 2003.

[62] Harold Ossher and Peri Tarr. Multidimensional separation of concerns and the hyperspace approach. In Mehmet Akşit, editor, *Software Architectures and Component Technology*, pages 293–324. Kluwer Academic Publishers, 2001.

[63] Karl R. Popper. *Alles Leben ist Problemlösen (über Erkenntnis, Geschichte und Politik)*. Piper Verlag München, 1994.

[64] Franz Prilmeier. AOP und evolution von software-systemen. Master's thesis, Technische Universität München, November 2004.

[65] A. Rashid, P. Sawyer, A. Moreira, and J. Araujo. Early Aspects: A Model for Aspect-Oriented Requirements Engineering. In *IEEE Joint International Conference on Requirements Engineering*, pages 199–202. IEEE Computer Society Press, 2002.

[66] Awais Rashid, Ana Moreira, and Joao Araujo. Modularisation and composition of aspectual requirements. In *AOSD '03: Proceedings of the 2nd international conference on Aspect-oriented software development*, pages 11–20, New York, NY, USA, 2003. ACM Press.

[67] Awais Rashid, Peter Sawyer, Ana Moreira, and Joao Araújo. Early aspects: A model for aspect-oriented requirements engineering. In *Proceedings of IEEE Joint International Conference on Requirements Engineering (RE 2002), IEEE Computer Society, pp. 199-202*, 2002.

[68] Martin Rinard, Alexandru Salcianu, and Suhabe Bugrara. A classification system and analysis for aspect-oriented programs. pages 147–158, 2004.

[69] Nathanael Schärli, Stéphane Ducasse, Oscar Nierstrasz, and Andrew P. Black. Traits: Composable units of behaviour. In Luca Cardelli, editor, *ECOOP*, volume 2743 of *Lecture Notes in Computer Science*, pages 248–274. Springer, 2003.

[70] Yannis Smaragdakis and Don Batory. Mixin layers: an object-oriented implementation technique for refinements and collaboration-based designs. *ACM Trans. Softw. Eng. Methodol.*, 11(2):215–255, 2002.

[71] Ian Sommerville. *Software Engineering*. Addison-Wesley, 7th edition, 2004.

[72] Zoltá Gendler Szabó. Compositionality. In Edward N. Zalta, editor, *The Stanford Encyclopedia of Philosophy*. Spring 2005.

[73] Peri Tarr and Harold Ossher. *Hyper/J User and Installation Manual.* IBM Research, 2000.

[74] Twente University TRESE Group. *Compose* Annotated Reference Manual*, October 2006.

[75] UMLsec tool, 2002-04. Open-source. Accessible at http://www.umlsec.org.

[76] Klaas van den Berg and José María Conejero. A conceptual formalization of crosscutting in AOSD. In *DSOA'2005 Iberian Workshop on Aspect Oriented Software Development*, Technical Report TR-24/05. University of Extremadura, 2005.

[77] Klaas van den Berg, José María Conejero, and Ruzanna Chitchyan. AOSD Ontology 1.0 - public ontology of aspect-orientation. AOSD-Europe-UT-01 D9, AOSD-Europe, May 2005.

[78] Klaas van den Berg, José María Conejero, and Juan Hernández. Analysis of crosscutting across software development phases based on traceability. In *Workshop in Aspect-Oriented Requirements Engineering and Architecture Design*, Shanghai, 2006.

[79] Cristina Videira-Lopes. AOP: A historical perspective. In R. Filman, T. Elrad, S. Clarke, and M. Akşit, editors, *Aspect-Oriented Software Development*, pages 97–122. Addison-Wesley, 2004.

[80] David Walker, Steve Zdancewic, and Jay Ligatti. A theory of aspects. In *ACM SIGPLAN International Conference on Functional Programming, Uppsala, Sweden, August*, 2003.

[81] Alfredo Weitzenfeld. *Ingeniería de Software Orientada a Objetos con UML, Java e Internet.* Thomson Learning Iberoamérica, México, 2004.

[82] Roel J. Wieringa. A survey of structured and object-oriented software specification methods and techniques. *ACM Comput. Surv.*, 30(4):459–527, 1998.